Advance Praise for *The Organic No-Till*

The Organic No-Till Farming Revolution will be a game-changer for flower farmers and so many other growers, big or small. The conversational style combined with practical and proven techniques make the no-till methods described in each case study very approachable and replicable. The book emphasizes the many environmental and economic benefits of no-till farming and the fact that you do not need expensive equipment to farm intensively, organically and profitably on a small scale. If you weren't already convinced about the many benefits of organic farming, this thoughtfully written book will undoubtedly persuade you.

—Erin Benzakein, author, *Floret Farm's Cut Flower Garden*

This book is likely the most practical examination of no-till farming methods since *Plowman's Folly*. Any farmer looking to reduce—or eliminate—tillage will find fresh ideas in these pages.

—Ben Hartman, author, *The Lean Farm* and *The Lean Farm Guide to Growing Vegetables*

Although long considered the gold standard of a sustainable and resilient agriculture, the many practical challenges of organic no-till have limited its use. Mefferd's new book shares a wealth of inspiring stories of innovative small-scale organic growers who have successfully overcome those challenges to reap the benefits of organic no-till on farm and in community.

—Laura Lengnick, author, *Resilient Agriculture*

Many of the young people I meet who would like to get into farming are discouraged because they don't have money for land and equipment. Andrew Mefferd's new book shows that you don't need a lot of money to get started. No-till farming doesn't require expensive equipment — and it's better for the environment. I hope the case studies he presents here will convince aspiring farmers and established farmers alike to reconsider the necessity of tilling. Let the no-till revolution begin!

—Lynn Byczynski, author, *The Flower Farmer* and *Market Farming Success*

The health of our soil is a major player in the success of a more resilient agriculture. *The Organic No-Till Farming Revolution* is full of practical advice to change the way we grow from the ground up.

—Zach Loeks, author, *The Permaculture Market Garden*

The only way to produce nutrient-dense food is with healthy soil. With *The Organic No-Till Farming Revolution*, Andrew Mefferd provides us the template to do just that while being highly profitable. I highly recommend this book.

—Gabe Brown, regenerative farmer, rancher and author

Here is actionable information for farmers who want to increase the amount of no-till growing on their small-scale farm. You don't have to invest in expensive cumbersome machinery or be an enthusiast of permanent no-till everywhere (which is difficult in organic farming) to benefit from some very practical new tricks. Different strategies work for different farms and different crops.

Andrew says in the introduction, "No-till is as much about climate change as it is about soil health as it is about farm profitability." Work on all three at once with these methods. The first part of the book explains the concepts. Mulch grown in place; applied cardboard, deep straw or compost; occultation (tarping) and solarization (clear plastic) are the options covered. The main part of the book consists of in-depth interviews with seventeen farmers about what works for them.

—Pam Dawling, Twin Oaks Community, Virginia, author of *Sustainable Market Farming* and *The Year-Round Hoophouse*

Inspiring and practical advice from the front lines of the soil-health revolution.

—David R. Montgomery, author, *Growing A Revolution: Bringing Our Soil Back To Life.*

Andrew has compiled an impressive number, and range, of actual farms making no-till work on their small farms. It's great to hear from so many farmers who are building healthy soil by replacing steel in their fields with organic matter and biology, and are reaping big harvests as a result.

—Josh Volk, slowhandfarm.com, author of *Compact Farms*

THE
ORGANIC
NO-TILL
FARMING
REVOLUTION

HIGH-PRODUCTION METHODS
FOR SMALL-SCALE FARMERS

ANDREW MEFFERD

Foreword by Kai Hoffman-Krull

new society
PUBLISHERS

Cover image: Digital composite illustration by Diane McIntosh: (using image elements) ©iStock 513708423, 521312440, 598560384, 668003964, 817298318, 827963920, 861537760, 862359710

All interior photographs © Andrew Mefferd 2019, unless otherwise noted; p. 1 © geraria; p. 53 © Viktoriya Sukhanova/Adobe Stock.

Printed in Canada. January 2019.

Inquiries regarding requests to reprint all or part of *The Organic No-Till Farming Revolution* should be addressed to New Society Publishers at the address below. To order directly from the publishers, please call toll-free (North America) 1-800-567-6772, or order online at www.newsociety.com

Any other inquiries can be directed by mail to:

New Society Publishers
P.O. Box 189, Gabriola Island, BC V0R 1X0, Canada
(250) 247-9737

LIBRARY AND ARCHIVES CANADA CATALOGUING IN PUBLICATION

Mefferd, Andrew, author
The organic no-till farming revolution : high-production
methods for small-scale farmers / Andrew Mefferd ; foreword by
Kai Hoffman-Krull.

Includes index.
Issued in print and electronic formats.
ISBN 978-0-86571-884-5 (softcover).—ISBN 978-1-55092-677-4 (PDF).—
ISBN 978-1-77142-272-7 (EPUB)

1. No-tillage—Handbooks, manuals, etc.   2. Organic farming—
Handbooks, manuals, etc.   3. Alternative agriculture—Handbooks,
manuals, etc.   4. Sustainable agriculture—Handbooks, manuals, etc.
5. Farms, Small—Handbooks, manuals, etc.   I. Hoffman-Krull, Kai,
writer of foreword   II. Title.

S604.M44 2019   631.5′814   C2018-906456-0
                                                       C2018-906457-9

Funded by the Government of Canada   Financé par le gouvernement du Canada

To Cleome and Jasper.

You are the future.

Contents

Foreword: The Age of Carbon

by Kai Hoffman-Krull

The modern age could very well be termed the age of carbon. We have increased the amount of carbon dioxide in the atmosphere by more than a third since the Industrial Revolution began.[1] A gas that keeps heat from the sun contained within the Earth's atmosphere, carbon dioxide makes up more than three-quarters of the greenhouse gas emissions in the world.[2] At the same time agriculture is currently experiencing a carbon crisis, with 50–70 percent of the world's carbon in farmland soils off-gassed into the atmosphere due to tillage.[3] Carbon, known as the building block of life, is the single most essential element in soil fertility as it aids in soil structure development, water retention, nutrient retention, and the biological process.

The decreased fertility from our carbon loss is occurring during a changing climate, when creating resilient crops that can withstand the stress of unpredictable weather patterns will be more important than ever before. The Intergovernmental Panel on Climate Change estimates that global food production could be reduced by up to 17 percent by the year 2100 due to crop failures from increased weather variation.[4] The population in the year 2100 is estimated to be 11.2 billion people.[5] Finding ways to preserve the carbon in our soil is simultaneously an environmental and social piece of activism, something we can do on our farms to improve our soil health and the health of our climate.

One of the most central carbon retention practices is no-till cultivation. Tillage has contributed 792 billion tons of carbon emissions over the past 250 years.[6] In comparison, humans contributed nearly 40 billion tons of carbon dioxide into the atmosphere last year. Tillage introduces unnaturally large amounts of oxygen into the soil, increasing the decomposition of organic matter. As carbon from this organic tissue is freed through the decomposition process, carbon molecules bond with the abundant oxygen introduced through tillage to become CO_2, rising into the atmosphere.

If you've used tillage and seen impressive results, that's because tillage is indeed providing a biological bloom momentarily in your soil. By increasing the decomposition of soil organic matter, there is a short-term rise of available labile carbon—the form of carbon that fuels the microbial machinery. While the fungal hyphae are torn and disrupted through tillage, this available labile carbon generates a rise in soil bacteria, which increases the percentage of nutrients that are bio-available for root uptake.

The problem is that tillage is mining this organic carbon at a very quick rate that provides immediate nutrient gain but at a significant long-term cost. A research colleague, Dr. Tom DeLuca at the University of Montana, found that tillage in Midwest prairie soils decreased organic matter levels by 50 percent over a fifteen-year period. The additional concern with this decreased organic matter is that soil carbon levels operate exponentially. Higher rates of organic matter allow for increased nutrients and water to be made available, which in turn increases the production for cover crop and green manure material—two of the foundational methods of increasing soil organics. With decreased organic matter levels, production of both market crops and cover crops decreases over time, making it more difficult to regenerate from the carbon deficiency created through tillage.

My farming mentor, Steve Bensel, once told me that almost everything we do in sustainable agriculture—cover cropping, animal rotations, reduced tillage, composting—are all fundamentally about increasing organic matter in the soil. And when we speak about organic matter we are in large part speaking about carbon, which comprises 58 percent of soil organic matter.[7] Organic matter and the carbon within it holds several key roles in soil health:

Microbiome

Carbon is the fuel source that drives the microbial network to digest minerals and make them bio-available to plant roots, also known as mineralization. Without this biological support system processing minerals, plants find it more difficult to access the nutrients available in the soil.

Soil Aggregation

The sugars from composted organic matter pull soil particles into aggregates, providing space that allows soil to store air and water. As this structure diminishes with tillage, soil compacts more and more, requiring higher amounts of disturbance for water, air, and roots to access the subsoil layers.

Water

Organic matter can absorb six times its weight in water, playing a significant role in holding moisture in the soil.[8] In addition, the decreased compaction of no-till plots allows for water access through the soil layers, whereas compacted soil creates runoff that carries water and nutrients away. In a four-year study at the University of Nebraska, researchers found that no-till plots saved between two-and-a-half to five inches (65–130 mm) of water per year compared to tilled plots.[9]

Nutrients

Organic matter increases the soil's cation exchange capacity, a measure of the soil's ability to hold nutrients. This means less fertilizer costs each growing season.

The No-Till Solution

No-till systems operate in a manner that mimics natural soil ecosystems—the microbiome, soil animals, and root fibers develop a lattice tunnel system that aerates the ground. Through limiting the loss of organic matter in the soil, no-till methods improve these key soil areas of biological activity, structure, and water and nutrient retention. Unlike tillage, which maximizes benefits in the short term while decreasing soil health over time, no-till systems mature in their fertility. No-till can regenerate compacted, disturbed soils and return carbon back to the ecosystem. No-till is not the only carbon solution we must explore to remediate our depleted croplands nationally and globally, but without it we should all fear for what our children and their children will eat.

If you start using no-till methods, tell your customers and friends. In a time when our government is actively removing environmental regulations, we need to find ways of inviting more of our populace to participate in climate solutions. No-till is a practice you can promote as increasing the quality of your produce, as well as storing carbon in the soil and keeping it from the atmosphere. You can see this as marketing, but also as environmental education—helping people understand the soil carbon crisis and ways they can participate in regenerating our farmlands through their purchasing decisions. And if you aren't a farmer, tell your co-ops and the farmers at the farmers market you want to buy no-till produce.

For as much as we all use the term sustainable agriculture, few of us contemplate the cost of which we are truly speaking—future famine. We have all lived through the peak of tillage agriculture, where food has been abundant as we have mined our soil resources to maximize immediate food production. Famine only exists for us in stories; it's something we read about in books, see in movies, or hear about occasionally somewhere else on the globe. Like climate change, it can feel like an abstraction. But famine may not be an abstraction to future generations. We cannot avoid the cost of what our food system has extracted, and some day that debt will need to be paid.

Copernicus started a revolution when he told us that the Earth was not the center of the universe. Today we need a new revolution, one where earth becomes the center of our human universe. We all eat. May this book, and you, be a part of that revolution.

—Kai Hoffman-Krull

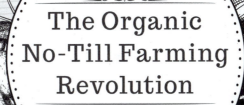

The Organic No-Till Farming Revolution

1

Introduction

Connections Between Soil Health, Climate Change, and Farm Profitability

Themes of soil health, climate change, and farm profitability came up over and over through the course of the interviews for this book. To the point you might be left wondering, is this book about no-till, or is it about climate change, or soil health, or small-farm profitability?

No-till is as much about climate change as it is about soil health as it is about farm profitability. No-till growing practices are a way to improve all three. As Kai Hoffman-Krull's foreword shows, we have to start farming more ecologically if we want to survive as a species. And small-farm profitability is important because no one will have a small farm if they can't make a living at it.

Ultimately, no-till is about the soil, and how improving soil health can also improve atmospheric health and farm bottom lines. Any one of these issues by itself is compelling enough to make us want to try no-till. The fact that no-till makes the connection between all three issues is what makes it so timely.

For example, if you only cared about farm profitability, and didn't care about soil or atmospheric health, no-till would still be worthwhile for improving farm efficiency and profitability. Growers who are happy with what they are earning, but want to grow in a more ecological method, will also be interested in no-till.

Mining the Soil, Polluting the Air

The conventional way of farming is contributing to the destruction of both our soils and our atmosphere. On the other hand, no-till farming practices build soil and sequester carbon at the same time. With conventional tillage, we've been mining the soil, breaking down soil organic matter (SOM) faster than it can be replaced.

In their book *The Hidden Half of Nature*, David R. Montgomery and Anne Biklé describe "one of the oldest problems plaguing humanity— how to grow food without depleting or destroying the soil.… By nurturing the microbial life below ground, we can reverse much of the damage caused by the ancient practice of plowing and the modern overuse of pesticides and fertilizers."[10]

How is tilling mining the soil and polluting the air? One of the reasons tillage is effective is because it speeds up the decomposition of organic matter (OM) in the soil. This reduces the amount of OM in the soil, and means that more carbon, which is stored in the OM, gets transferred to the atmosphere as CO_2 instead. So as we are hurting our soils, we are also hurting our atmosphere.

This is a terrible situation to be in, with soil becoming so depleted and CO_2 levels so inflated that they are threatening to drastically change the planet, at best, and at worst make it unlivable. The only good news here is that the no-till practices that help the soil also help the atmosphere. So as farms adopt no-till practices that make their soil healthier, it should also help their profits and the atmosphere at the same time.

A significant amount of the Earth's surface is taken up with farming, so turning it into a carbon-building exercise instead of a carbon-releasing one would make a big difference.

Thanks to the Growers

Without them there would be no book. I love farm tours so to be able to spend the year visiting with growers pioneering systems that have the potential to revolutionize agriculture led to a magnificent journey that I'm excited to share with you. Farm after farm, I was inspired by who

I met and what they were doing, only to move on to the next farm and be inspired once again.

I hope this book conveys the excitement I experienced meeting these growers and seeing their methods. It was particularly exciting to see how many of the growers in the book cited other growers they hadn't even met as inspiration, having learned from others through writing, conferences, and social media. I hope this book furthers that process. For taking the time to meet with me during the busyness of the season, I am grateful to the growers for sharing their time and methods with us.

The Importance of Lowering Barriers to Beginning Farmers

With such a small percentage of the population in farming, the only way for us to have a more resilient, healthier food system is to get more people farming again. That is going to involve a lot of people who weren't born to it getting into farming.

"When less than two percent of the population is producing the fundamental nourishment for the rest, it requires chemical and industrial methods that are depleting the soil, polluting the water, and making us sick," said Michael Abelman in the foreword to Josh Volk's excellent book *Compact Farms*.

We need to make it easier for people to start farms. No-till complements the ideas in *Compact Farms*, by saying, not only do growers not need a lot of land to run a commercial farm, they may not even need to invest in machinery.

The thing that needs to happen to keep people in farming is that small farming has to be profitable—so not just the determined will do it for a livelihood. That's why the examples in this book give me hope. There are numerous examples of farms that were able to start small with minimal mechanization, and grow as their businesses grew and make a decent living.

I'm especially excited by the potential of no-till to encourage people to give farming a try. The barriers to starting a farm are high, including access to land and equipment. No-till makes it possible to start a farm without a tractor or even a rototiller. I have visions of kids no-tilling up their parents' suburban yards (ask permission first, kids!), city growers making the most of vacant lots, and rural growers no-tilling whatever land they have.

How to Use This Book

The first part of the book is the quick start guide. The interviews are the detail and the supporting material. The intro is written to answer the question of what organic no-till systems are, why they were developed, and which methods should be used in which situations. The majority of the book (the interviews) show the details of how people are making these systems work with a variety of environments and crops.

I don't imagine most people reading this book straight through. I'm guessing that after looking through the methods, one (or more) will stand out from the others. Then I imagine people skipping to the interviews covering their method of interest, to decide if they want to try a particular method. Without adequate information, most people who are interested will not take the step of trying the methods. In this manner I hope to transfer knowledge from the practitioners to those who are interested, and help promote the organic no-till revolution.

It's important to note that some of the growers have their own sources of information. Conor Crickmore of Neversink Farm has extensive online courses available on his website. Bryan O'Hara of Tobacco Road Farm has a forthcoming book which I look forward to reading. Paul and Elizabeth Kaiser of Singing Frogs Farm have a number of Youtube videos on their philosophy and methods. Ricky Baruc of Seeds of Solidarity has a Youtube video and does workshops at his farm. Tony and Denise Gaetz of Bare Mountain Farm have extensive materials on Youtube and their website. Shawn Jadrnicek of Wild Hope Farm talks about his no-till methods in his excellent book, *The Bio-Integrated Farm*. And I look forward to covering the evolution of no-till in *Growing for Market* magazine. See the Resources section of the book for more details on how to find these materials.

If you already know you want to try no-till, you might want to skip straight to the methods section, and then on to the interviews. If you're looking to understand what no-till methods are and why they're important, start at the front of the book.

Who This Book Is For

Have you ever wished you could use less machinery on your farm, and still be highly efficient and productive? Do you want to start a viable commercial farm that will pay a living wage, with minimal investment in equipment and land? Do you have a small piece of land and are wondering if it can be a commercial farm? Want to build organic matter and soil biology because of the way you grow, instead of in spite of it?

The benefits of no-till sound almost too good to be true. In this book, read about the farmers who use these systems to run profitable commercial farms, and decide if one of them is right for you.

Whether your goal is to spend less time on a tractor, burn less fossil fuel, own less equipment, be more efficient and save labor, sequester carbon, or build soil, no-till farming methods can contribute to all of these goals at the same time.

This book is for people who are getting into farming and considering what system to use. It is also for people who have wondered how they might simplify their existing systems.

What I want to make clear is that I'm a promoter, not a proselytizer. This isn't like a religion where I'm trying to get everybody to convert to doing the same things. Over the course of traveling and writing this book, in addition to no-till enthusiasts and skeptics, I've encountered many good growers who rely on tillage and are happy with their systems. I would say: If you're happy with your system, keep doing what you're doing. I'm not trying to talk people who have spent years dialing-in their systems to abandon them.

This is written for growers who are not completely happy with their tillage systems, or new growers who are thinking about starting a farm and want to take no-till methods into consideration. If this book makes growers think about the benefits of tilling less, whether they go all the way to no-till or not, then it will have done its job.

Reducing the amount of tillage has benefits for farms of all sizes. This book is about organic no-till solutions for small growers, because the conventional no-till solutions are not compatible with organic farming. And some of the practices that work on larger organic farms don't work well for smaller farms.

I present this collection of interviews in order to offer solutions for smaller growers, because they have so much to gain from them. May it help you farm more simply, more efficiently, more profitably, and more ecologically.

The Title

No-till has the potential to be a farming revolution. Tillage has historically been such a dominant paradigm that organic no-till is disruptive technology for small farms. If you could farm without tillage, why would you keep tilling?

It wasn't until we were deep into the book process that I noticed the similarity in titles with *The One-Straw Revolution*, a book by the Japanese farmer and philosopher Masanobu Fukuoka. The similarity was not intentional but is significant because, during the interviews, multiple growers brought up *The One-Straw Revolution* as a source of inspiration.

Though growers have had difficulty directly applying Fukuoka's ideas here in North America, the current wave of no-till continues in the same spirit. As Wendell Berry wrote in his preface to *The One-Straw Revolution*, "Knowledgeable readers will be aware that Mr. Fukuoka's techniques will not be directly applicable to most American farms. But it would be a mistake to assume that the practical passages of this book are worthless to us for that reason."

The no-till methods as explained by their practitioners continue in Fukuoka's footsteps in order to farm as much as possible with natural systems.

Growing a Revolution

Another book with the word "revolution" in the title that I took a lot of inspiration from was *Growing a Revolution: Bringing Our Soil Back to Life* by David R. Montgomery. Here I want to quote a long passage from the book because I think it puts our current perilous agricultural and human survival situation in historical perspective. But first, I want to mention that Montgomery wrote another book, called *Dirt: The Erosion of Civilizations*. The jacket description says that *Dirt* "blends natural and cultural history to show how soil erosion caused past civilizations to crumble and how modern agricultural societies face a similar fate unless they shift to more sustainable practices." So when he talks about civilizations that collapsed due to the destruction of their soil, the guy knows what he's talking about.

A section of *Growing a Revolution* called "A New Revolution" begins:

A look back at our agricultural past reveals a long series of innovations, and a few bona fide revolutions, that greatly reduce the amount of land it takes to feed a person. These changes led to a dramatic increase in how many people the land can support and a corresponding decrease in the proportion of people who farm. By my reckoning, we've already experienced four major revolutions in agriculture albeit at different times in different regions.

The first was the initial idea of cultivation and the subsequent introduction of the plow and animal labor. This allowed sedentary villages to coalesce and grow into city-states and eventually sprawling empires. The second began at different points in history around the world, as farmers adopted soil husbandry to improve their land. Chiefly, this meant rotating crops, intercropping with legumes (plants that add nitrogen to soil), and adding manure to retain or enhance soil fertility. In Europe this helped fuel changes in land tenure that pushed peasants into cities just in time to provide a ready supply of cheap urban labor to fuel the Industrial Revolution.

Agriculture's third revolution—mechanization and industrialization—upended such practices and ushered in dependence on cheap fossil fuels and fertilizer intensive methods. Chemical fertilizers replaced organic matter-rich mineral soil as the foundation of fertility. Although this increased crop yields from already degraded fields, it took more money and required more capital to farm. This, in turn, promoted the growth of larger farms and accelerated the exodus of families from rural to urban areas. The fourth revolution saw the technological advances behind what came to be known as the green revolution and biotechnology breakthroughs that boosted yields and consolidated corporate control of the food system through proprietary seeds, agrochemical products, and commodity crop distribution—the foundation of conventional agriculture today.

What will the future hold as we burn through the supply of cheap oil and our population continues to rise alongside ongoing soil loss and climate change? A recent study authored by hundreds of scientists from around the world concluded that modern agricultural practices must change once again if society is to avoid calamitous food shortages later this century. Just how worried should we be? Well, consider the fate of Mesopotamia, ancient Greece, or other once-great civilizations undone by their failing land. This time we need to ask what agriculture would look like if we relied on building fertile soil instead of depending on chemical substitutes. What would this new, fifth agricultural revolution look like?[11]

Note that the first agricultural revolution was cultivation. Montgomery goes on to point out that there are different methods that offer us the "opportunity to break free from the cycle of land degradation that doomed ancient societies."

We are going to need everything we've got to break the cycle of land degradation and escape the environmental doom that faces our own society. No-till can be part of the remedy for the agricultural and environmental problems made worse by tillage that threaten us. The fifth agricultural revolution has to be about undoing the damage from the previous four, if we have any hope of not joining the club of "other once-great civilizations undone by their failing land," as Montgomery calls them.

Life on the Edge of Collapse:
Sharpening the Axe to Cut Down the Last Tree

As *Dirt* shows, it's not a new thing for human societies to collapse due to environmental degradation. The novel difference in the situation we face is that we have environmental damage on not just a civilizational but a global scale (agricultural soil loss, climate change, pollution, etc.) that threatens all of the 7.6 billion and growing people on Earth.

When I hear population projections like the one in the foreword to this book—that we'll have 11.2 billion people on the Earth by the end of the century—I am skeptical we'll ever reach those numbers. At some point the human population will exceed the carrying capacity of the planet.

Just as plants' growth is limited by not having enough of just one nutrient, if human population exceeds the carrying capacity of the Earth in any single aspect (e.g., amount and quality of agricultural soils, livable climate, etc.) then it will have to stop growing. In that no-till sequesters carbon and builds soil, it can help undo some of the damage to our atmosphere and our soil at the same time.

I first learned the history of Easter Island in Jared Diamond's excellent book *Collapse*. It is an isolated sixty-square-mile island in the Pacific. The very abbreviated version of its story is that the island's population crashed from around fifteen to just two to three thousand people over the course of a century due to environmental degradation including deforestation and invasive species, among other causes.

Montgomery also looks at Easter Island in *Dirt*. Famous for its giant stone heads erected during more prosperous times, "Easter Island presented a world-class puzzle to Europeans who wondered how a few stranded cannibals could have erected all those massive heads. The question mystified visitors until archaeologists pieced together the environmental history of the island to learn how a sophisticated society descended into barbarism. Today Easter Island's story provides a striking historical parable of how environmental degradation can destroy a society."

Thinking about Easter Island, I wondered how someone could possibly have cut down the last tree. Even if they didn't know for sure it was the very last one, on such a small island they must have noticed they were getting low on trees. It boggles the mind how the people didn't make the connection between elimination of trees and their own survival.

But then I look at our current situation and once again see a population on the edge of collapse due to environmental degradation, only this time on a global scale. I would like to think that our superior technology would let us see the problem of climate change in time to take action. However at this point we seem to be in the same position as the Easter Islanders; too stuck in the way we are used to doing things to change in time to avert disaster.

One of the points Jared Diamond makes in *Collapse* is that societies often collapse shortly after their peak, because they're peaking as they're outstripping their resource base. To apply the principle to the present, as we outstrip our global resource base and overgrow our environmental carrying capacity, the present day will be viewed as the peak before a scarcity-of-resources-induced collapse unless we figure out how to solve the enormous problems we've made for ourselves.

I imagine there's a certain type of hubris in a peaking society that obscures the coming fall, which we are currently prone to. Living at the pinnacle of human potential dazzles us with what we have achieved, to the point where we think we're invincible. Something along the lines of, "I've got a computer in my pocket. I'm sure if global warming gets really bad, we'll be able to fix it." Or, "We've got self-driving cars, how could starvation be caused by anything as mundane as the degradation of our agricultural soils?"

That is the deception in the often-heard argument by chemical agriculture proponents—that organic won't feed the world. It's an argument that works because it plays into the notion of the techno fix: "Sure, we can have any number of people on the Earth. As long as we have enough people doing science we'll find a way to feed them!"

Whereas so few people have any connection to agriculture anymore they don't realize that all of the chemicals we are raining down on our crops are actually degrading the capacity of the land to support life. This is why we need solutions like no-till—simple, accessible actions that anyone can take to produce their own or someone else's food more sustainably.

Collectively, we are that person on Easter Island, poised to cut down our last tree. I often worry that last tree has already fallen, and that climate change and environmental degradation are already past the point of no return. But we don't know that for sure, so we must do what we can to try to reverse the damage we have caused.

Have we passed a tipping point on global warming? Have we cut down the last tree already? For better or for worse, it's impossible to know. The Earth is a lot bigger than Easter Island, and it's a lot more difficult to assess whether we've passed a global tipping point on climate change.

The fact that we are even asking the question of whether we have passed the tipping point for life on Earth should be terrifying to everyone on the planet. Unfortunately for us, humans aren't very good at dealing with enormous, slow-moving problems like climate change. If we knew aliens were on the way to raise the temperature to cook all of us off of the Earth, we would be marshaling all resources in a WWII-style mobilization to defeat the invaders. But cooking ourselves off the planet seems harder to get our arms around. We get bogged down in day-to-day matters of survival today instead taking action to save ourselves tomorrow.

I can't tell you that every vegetable grower going no-till would stave off the sixth mass extinction. My passion and expertise lie in agriculture, so I look for solutions to problems in my chosen field. Even though veggie growers going no-till won't change things all by itself, veggie growers operating more sustainably, plus a lot of other changes are the only shot we've got to keep times from getting extremely tough for ourselves in the future.

Against this backdrop, there is not a lot of good news. One of the only bright spots, also from *Dirt*, is another island in the Pacific Ocean, very similar in many ways to Easter Island, that was able to recognize its coming ecological collapse and avert it: Tikopia. "After seven centuries on the island, the islanders intensified pig production, apparently to compensate for loss of birds, mollusks, and fish. Then instead of following the path taken by the Mangaians and Easter Islanders, Tikopians adopted a very different approach," writes Montgomery.

Realizing that their environment was losing the capacity to support them, "Tikopians began adapting their agricultural strategy.... Over many generations, Tikopians turned their world into a giant garden with an overstory of coconut and breadfruit trees and an understory of yams and giant swamp taro. Around the end of the sixteenth century, the island's chiefs banished pigs from their world because they damaged the all-important gardens.

"In addition to their islandwide system of multistory orchards and fields, social adaptations sustained the Tikopian economy. Most important, the islanders' religious ideology preached zero population growth."[12]

The connection between our resource base and our population is one we seem to be having trouble making as a species. At the very least we need to be like the Tikopians and stop destroying our resource base.

The fact that the ideas in this book also contribute to farm efficiency, profitability, and lowering barriers to starting farms is what gives me the hope that they will be adopted on a large scale. No-till is one of those solutions that is better for the planet and the bottom line, which means it's more likely to happen. Because doing the right thing is much more likely when it also makes life more efficient and profitable. It is with this in mind that I write, hoping this book will be my own little contribution to the healing of the world, the climate, and indeed to having a future at all.

I am reminded of another lesson from *Collapse*—the genocide in Rwanda was partially caused by overpopulation. Overpopulation can result from having more people than a given environment can support, or it can result from the degradation of the network that has otherwise comfortably supported people.

The question that pains me the most is not whether we survive as a species or not; resource scarcity is the last thing I want to imagine my kids having to live through. Yet every day I am confronted with the prospect that I brought children into the world only for them to see it fall apart. With a growing population's chances of feeding itself progressively diminished by both destruction of good agricultural soil and climate change, it's very difficult for me to imagine that we will not have to deal with painful resource scarcity within my lifetime.

For it's not just starving to death that we need to fear. The very real question of who dies and who lives through a period of resource scarcity is an unpleasant one to resolve. I'll direct you back to Jared Diamond's *Collapse* for a more recent example of the type of suffering that occurs during such a situation. He shows how many of the societal collapses through history have been in part due to loss of soil through erosion, salinization, or loss of fertility. Eras of resource contraction are not pleasant times to live in.

The only way for us not to become Easter Island on a global scale is to take better care of our planet. We have to save our own world. Let's get started.

How I Got Interested in No-Till, Lost It for a Decade, and Found It Again

In 2004, I worked on a 100-acre organic vegetable farm on the West Coast. Because herbicides weren't an option, and black plastic mulch wasn't used, cultivation was constant. I learned how to drive a tractor really well.

This left me thinking, there's got to be a better way to keep weeds down than cultivation. I started hearing about no-till practices as a way to get rid of the ills of tillage and weeds at the same time, but there wasn't much actionable information out there. So I started researching it.

If you looked up "no-till" online in 2004, a lot of the references were to the roller-crimper style no-till (see Biodegradable Mulch Grown in Place, p. 36). I found information from the Rodale Institute, the

USDA, and a professor at Virginia Polytechnic Institute and State University [Virginia Tech] named Ron Morse, among others.

At the time I was apprenticing on farms over the summer and coming home to Virginia to a job that would take me back every winter in order to save some money to start a farm. Since I was headed back home to Virginia anyway, I got ahold of Ron and asked him if I could come down and pick his brain about no-till.

At some point in the winter of 2004–05 I found myself in Blacksburg, VA, at Virginia Tech's Kentland Research Farm talking with Ron Morse about his work with no-till. He offered me a job and I ended up working the 2005 farming season with Ron and his graduate student Brinkley Benson. Ron and Brinkley would design grants to explore questions related to organic no-till vegetable production, and it was my job to work with Brinkley to carry out the fieldwork for the grants. As an example, a lot of the grants would be something along the lines of comparing the inputs and productivity of organic no-tilled broccoli vs. organic clean-cultivated broccoli.

The transplanter that was modified to be no-till we used at Virginia Tech: A tank in front held water that was dribbled into the furrow that was cut through crimped cover crop residue by the large straight coulter. The shank behind the coulter loosens soil, and the boxes drop solid fertilizer into the furrow, metered by a chain attached to the wheels. The big black spools hold drip tape. Finally, two wooden seats are at the end where the two white bins hold transplants for the people to put in the transplanter.

Credit: Andrew Mefferd

It was an honor and a pleasure to work with both Ron and Brinkley. I look back and think about how lucky I was to end up working with two such talented agriculturalists, and am grateful for the experience. Brinkley and I worked together on a daily basis, and I still often think of one thing he would say when our progress was much too slow on some task: "We've got to find another [faster] gear."

Over my time working at Virginia Tech, I saw the roller-crimper method work really well. The next year I started a farm by leasing three acres from my grandmother in Pennsylvania. What I realized when trying to apply roller-crimper no-till to a three-acre market garden is the method is more suited to larger plantings of crops, like a field of sweet corn or tomatoes, a patch of pumpkins or squash, or other space-extensive plantings like field crops.

In the thick of starting a farm, I forgot about no-till when I realized the roller-crimper method was not well suited to the farm I was starting. I fell back on the more conventional tillage methods I had learned working on other farms: some combination of moldboard plowing, discing, harrowing, rototilling, and clean cultivation to develop a plantable seedbed and deal with weeds.

I continued on in this manner for a little over a decade until the winter of 2016–17. A few things happened that year that reinvigorated my interest in no-till.

First, in May of 2016, *Growing for Market* magazine ran an article by Jane Tanner, "The Many Benefits of No-Till Farming," that was an overview of market farm no-till techniques, profiling Neversink, Spring Forth, Four Winds, Bare Mountain, and Foundation Farms.

Then in January of 2017 I saw Paul and Elizabeth Kaiser of Singing Frogs Farm speak at the NOFA–Massachusetts conference about no-till on their farm. The next month I saw Bryan O'Hara of Tobacco Road Farm speak about his no-till methods at the NOFA–VT conference.

I realized that these growers had figured it out. There was a critical mass of people who had developed ways of doing what I had hoped to do a decade before: run a small farm without tilling.

I wanted to put their methods to use on my own farm. I went to see what information I could find about putting them into practice. When I couldn't find more than scattered information about what people were doing for no-till on a small scale, it crystallized the idea for this book.

I wanted to answer two questions: Will this work for me, and if so, how do I do it? I wanted to prevent others from finding themselves in the same situation I had, of having learned a no-till method only to find that it wasn't compatible with their farm.

Since people were using a number of different methods, and no one person was using all of them, I knew what I needed to do was go visit as many of them as possible and write up the interviews to guide and encourage people wanting to get started with organic no-till on a small scale.

The fact that there is more than one way to do most agricultural jobs is one thing that keeps farming interesting for me. There are as many ways to farm as there are farmers. Certain methods may work better than others on any given farm, not to mention different growers' styles and preferences.

I wanted to see for myself all the no-till methods that were working on farms. I wanted to survey what people were doing, their successes and their struggles, in order to pass on the information and let growers decide for themselves which methods to use. In some cases, individual growers have their own materials that may be more in-depth than this book. See the Resources section for a directory of the individual growers' information.

In addition to sharing this information with other growers, I wanted to finally get back to what I was trying to do in the first place, and use the information I gathered to decide on the best system(s) to implement no-till on my own farm. This book is as much for myself as everyone else. I want to reconfigure my farm and complete the journey I started 15 years ago.

2

Understanding
No-Till Systems

Drive around any rural area in the springtime and you're likely to see
freshly tilled fields being made ready for crops. Tillage is so basic to
agriculture it's a paradigm that is frequently not questioned. We can't
understand no-till systems and why they are advantageous until we
put them in the context of tillage and the disadvantages that go along
with it.

The Disadvantages of Tillage

Tillage is one of the most time, labor, and equipment intensive tasks on
the farm. It's easy to see that a lot of time and effort could be saved if
tillage were eliminated. The problem has always been how to prepare
the soil for planting without tillage?

"Tilling the soil is the equivalent of an earthquake, hurricane, tornado, and forest fire occurring simultaneously to the world of soil
organisms." What radical, tillage-hating group made such a strong
statement? The USDA-NRCS, in a pamphlet entitled "Farming in the
21st Century: A Practical Approach to Soil Health."

It goes on to say, "Physical soil disturbance, such as tillage with a
plow, disk, or chisel plow, that results in bare or compacted soil is destructive and disruptive to soil microbes and creates a hostile, instead
of hospitable, place for them to live and work. Simply stated, tillage is
bad for the soil."

Tillage results in two self-perpetuating cycles: it burns up soil
OM necessitating the addition of more, and it stirs up weed seeds,

necessitating yet more tillage to kill the weeds. Conventional farming "solves" these two problems in a manner that is not sustainable. For depletion of organic matter, it treats the soil as a substrate for holding plants and disregards the depletion of OM. For weeds, it has herbicides.

Organic agriculture offers improvements over conventional bare tillage. Most notably, organic system plans mandate that cover crops be grown between cash crops in order to add some organic matter back to the soil, and to keep the soil covered when it is prone to erosion (over the winter, for example).

Soil has three properties that we are most interested in agriculturally: the physical, the biological, and the chemical. Tillage is bad for all three of them.

On the physical side, the action of tilling crushes the soil structure, making soil more likely to erode and less likely to absorb water efficiently. On the biological side, the action of tilling kills many of the organisms that live in the soil. Tillage breaks apart soil fungi and the aggregates they make that help soil resist erosion and promote water infiltration. Over time, this promotes a soil environment with more bacteria and less fungi.

And on the chemical side, accelerating the oxidization of organic matter promotes a short-term release of fertility, at the expense of the long-term reserves in the soil. Furthermore, the destruction of soil organic matter releases carbon that has been sequestered in the soil into the atmosphere as carbon dioxide.

In addition to the negative effects on the soil, tillage also wastes a lot of time and energy. On my farm I've often thought, "If we didn't have to spend all this time and energy tilling, we'd save a lot of time and energy."

Tillage ties up a lot of money, in the form of fuel, labor, and equipment. It also ties up a lot of time, both in the sense of the time that it takes to do the tillage, but also in the sense that other farm operations may be delayed due to tillage. For example, tillage can't be done when it's too wet or too dry, so farmers often find themselves waiting for the soil to dry out in the springtime to till, when the temperature is otherwise adequate to plant. If there was a cover crop on the ground before tillage, then you have to wait at least an additional two weeks for it to break down after tilling before planting.

No-till trades tillage for other methods of field preparation that are less complex, strenuous, and time-consuming. It is a less invasive, more efficient, and more profitable field prep process that grows healthy soil in order to grow healthy crops.

According to a USDA fact sheet, "A simple definition of soil health is the capacity of a soil to function. How well is your soil functioning to infiltrate water and cycle nutrients to support growing plants?"[13]

The two best understood areas of the soil are its physical and chemical properties. It has long been known that the physical condition and chemistry of the soil have a lot to do with the success or failure of crops. Now we know that the biology is very important too, but we still have a lot to learn about the biology of the soil.

Maybe it's because soil biology was not thought important that conventional systems were designed to operate in spite of whether the soil was healthy or not. Tillage implements crush the soil into plantable submission, chemicals kill anything that might compete with the crop, and chemical fertilizers replace the fertility that was either lost from the soil or was no longer being cycled efficiently by biology. The cumulative effects of these practices are erosion, loss of fertility, and dead, nonfunctioning soil.

Once again, organic systems do better by incorporating cover crops to make up for organic matter destroyed through tillage, and at least by not using all those chemicals. But I have come to think of most tillage systems as having built-in remedies to try and deal with the destruction that they cause.

Conventional systems try to get around degraded soil biology and physics by using chemicals to keep plants productive. Organic tillage-based systems try to promote the biology in spite of the damage they are doing to the soil. Over the course of doing the interviews for this book, I've come to think of no-till systems as operating because of soil biology, not in spite of it.

No-till systems have advantages when it comes to promoting a healthy soil system. For one thing, they're not burning up the soil OM through tillage in the first place, so they don't have to do the one step forwards/one step backwards dance of tilling and then adding more OM to make up for the tillage you just did. So we can say they make it easier to raise the percentage of OM on your soil test.

In addition to sequestering carbon, increasing OM improves all three aspects of the soil. Higher OM increases the tilth of the soil (physical) and the life in the soil (biological), which will in turn improve the availability of nutrients in the soil (chemical).

Over the course of these interviews I've come to see what we used to regard as the least important element of healthy soil as the most important. Let's go back to our simple definition of soil health as the capacity of a soil to function. In a healthy soil the biology can improve the physical and chemical properties. Thus the organic adage to "feed the soil to feed the crop." In no-till systems, I've seen how the biology is promoted specifically to make the soil texture and chemistry good for growing crops. Good biology builds good soil texture and chemistry.

In conventional systems, the opposite is true. Chemistry is used to make up for poor soil biology, texture, and chemistry. I think some of the systems in this book make the almost perfect "closed loop" system a lot of organic growers are looking for. Four Winds Farm, for example, has been using their system for two decades. They make their own compost, and over time the OM level in their soil has grown high enough that, even though the compost is not particularly high in fertility (by density or fertilizer standards), enough is being made available by biology to feed the plants.

More recently we've begun to understand the importance of biology—that it helps cycle nutrients in the soil and develop the aggregation that prevents soil from eroding. But soil biology is an area where we still have a lot to learn.

Let's go back to our definition of soil health as the capacity of a soil to function, so unhealthy soils are not very functional and healthy soils are highly functional. What chemical agriculture does is make unhealthy, low-functioning soils grow plants with quick hits of chemicals. Which is why conventional agriculture is compared to drug addiction—you're constantly adding more chemicals to make up for the damage of the previous chemicals, and constantly tilling (or spraying) to kill the weeds whose seeds you churned up the last time you tilled. Tilling more to make up for tilling. Spraying more to make up for spraying.

We need to start thinking of the health of the soil just like we think of the health of a forest, a field, a lake, or even a human community. These all can be self-sustaining ecosystems, with producers, predators, prey, and organisms that sustain them from season to season. In natural systems, or naturally managed agricultural systems, the soil can function as a self-sustaining community. But with frequent tillage, introducing the effects of "an earthquake, hurricane, tornado, and forest fire occurring simultaneously" on the soil ecosystem has the same effects such a cataclysm would have on a human community; not everything dies, but the larger organisms and fungi are disproportionally vulnerable when the physical environment of the soil is destroyed.

Repeated tillage has the same effect on the soil that repeated cataclysms would have on a human community; it throws the soil community into a cycle of constantly being destroyed and rebuilt, favoring the bacteria that survive to feed on those killed by tillage.

The Benefits of No-Till

Against the drawbacks of tillage we can evaluate the advantages of foregoing tillage. One of the most exciting things about no-till is that, if you already have a farm, you may not need to buy anything or only make a minimal investment to try the methods. Most growers already have what they need to try no-till lying around the farm.

Increased Efficiency of Time

Most tillage systems require at least three passes over the field before they are ready to plant, requiring no less than three different pieces of equipment, and a tractor or horses to pull them with. The no-till systems in this book typically skip the step of tillage by using a mulch that is either left in place or removed to prepare the soil for planting. These mulches require less investment than tillage in every aspect:

- No-till takes less time than tillage
- No-till takes less equipment than tillage
- No-till takes less energy (in the form of tractor or horse power)
- No-till doesn't burn up organic matter the way tillage does
- No-till should require less work to prepare a field than tillage, with an additional advantage. Tractor work has to take place when a field is sufficiently dry, meaning that in humid regions farmers are at the mercy of the weather to start getting their fields ready in the spring.

On paper this may not seem like a big deal but in practice, getting a late start to the season can have a real impact on profitability and happiness. In a wet spring, farmers are at the mercy of the weather, waiting for fields to dry out. It can be really frustrating to sit and wait as planting dates go by on the calendar and transplants get too big in the greenhouse because the field is too wet to till.

The transplanter being made ready for a day of transplanting broccoli.

Credit: Andrew Mefferd

Consider that most no-till systems require no tractor implements, and no tractor. The basic requirements are to smother whatever is growing in a field with some type of mulch (see the individual methods for specifics), fertilize, and plant. While tractors can be handy for moving things around, they are by no means necessary, and several of the no-till farms I visited didn't even have any.

Increasing the Viability of Smaller Farms

In a world where large-scale commodity agriculture is given so many advantages through subsidies and other government support, regenerative agriculture needs all the help it can get.

People want real food, as shown by the steady growth of farmers markets and organic food sales. The connection between having smaller farms and having more real food available may not at first be apparent, so let's make it clear: some farm models have a certain size below which they don't make sense. The size and expense of the infrastructure dictate the expanse of the farm.

No-till stops the equipment from dictating the scale of the farm and lets the farm be the size it wants to be. For example, it doesn't make sense to buy a $250K tractor to cultivate an acre. It doesn't even make sense to buy a $25K tractor to cultivate an acre. To cite a personal example, even though we started our farm on three acres, I felt like we needed a tractor, mainly for tillage. So we ended up buying a tractor for our three acres, and it always felt like a bull in a china shop working on our vegetable beds.

Not everyone wants to have a big farm. I want people to feel like they can start a farm whether they have access to a lot of land or money or not. More farms will be started if people can start them on very small acreages with very little investment. Then those who are successful can choose whether to scale up or stay small.

Credit: Andrew Mefferd

Filled with broccoli transplants.

I don't personally have any favoritism about farm size. I think that to increase the amount and access to local food, we need lots more small, medium, and large farms. Realistically though, as someone who wants to see this change take place, I know more people have the resources and management skills to start a small farm than a big one. Lowering the amount of investment and land needed to start a farm is an important way to get more farms started.

Down-Scalability

Because of the reduced requirement for equipment, no-till enables smaller units of land to be economically viable units.

This is especially important for urban and suburban farms. Most suburban and especially urban areas do not have large uninterrupted tracts of land, and land prices may be high. Reducing the size of the piece of land necessary for a viable commercial farm makes it possible for people to start successful farms on smaller pieces of land. In order to re-localize our food system, we need to have lots of new farms of all sizes everywhere—including where much of the population is concentrated, in cities and suburbs.

Taking a break from transplanting.

Credit: Andrew Mefferd

Also, a smaller "entry level" size for commercial farms will open up farming to more people. I've known a number of people who wanted to farm but could not afford the investment in land or equipment. Reducing the necessary footprint size increases the number of people who can start farms. And more small farms means more farms of all sizes. Because some people who start small farms will scale them up to medium and large farms. And what we need is more farms of all sizes everywhere.

As in any industry, economy of scale is often used to increase efficiency in farming. Alas, by definition economies of scale are not available to small farms. The cumulative effect of the efficiencies of no-till enables one to run a commercially viable, living-wage farm on a very small footprint. So a farmer doesn't feel like they have to grow on a large acreage in order to make a living.

Increased Efficiency of Organic Matter

The no-till growers I visited with saw OM levels go up quickly after adopting no-till practices. Increasing OM in the soil makes plants grow better for a lot of different reasons, so it is a best practice of farming to try and increase OM over time. Cover cropping and adding compost are best practices because in addition to adding nutrients to the soil, they tend to increase soil OM.

In addition to grinding and incorporating whatever is growing in the soil where a crop needs to go, the churning of tillage burns up OM and in the process releases nutrients. So tillage systems need to add OM every year in order to make up for what they burn up during tillage, just to maintain equilibrium and stay at a constant level. Which is why no-till growers see a rapid rise in OM after adopting no-till methods—they are building soil without the burning up of OM that occurs during tillage.

Simplicity

The beauty of these methods is their simplicity; some of them could be explained in a sentence. Deep mulching with compost, for example, could be boiled down to: Apply a thick enough layer of weed-free compost to suppress weeds, and then plant into it. Of course, more information than that is helpful to get started, since the devil (and the success) is in the details. That's why there are summaries of the methods later on in the introduction, and the real nitty-gritty details in the interviews.

One advantage: It's not rocket science! Really, the only reason we need a whole book about it is to cover all the different methods.

Reduced Mechanization

Along with their simplicity, no-till methods should result in a reduction in mechanization and the complications that go along with it: owning equipment, fixing it, fueling it, and the emissions it produces. I used to think that tractors were a necessary evil, but no-till made me realize they're not necessary for having a profitable farm.

Efficient Use of Space

With less space devoted to paths, turnarounds, and headlands for equipment, farms can be more productive because more of the space is devoted to growing crops. For most of those systems using permanent beds, fertility can be concentrated on the growing area where it is needed. Seeds can be scattered at higher density than with cultivation because space doesn't need to be left open for passes of the cultivator.

Quick Successions

Because time doesn't have to be taken to till between crops, many no-tillers I talked to were able to re-plant a harvested bed within the same

Making sure everything is working right. Once we got going, we could transplant very quickly into a high-residue bed that crimping produces.

Credit: Andrew Mefferd

day or very quickly after harvest. This maximizes the profitability and quick turnover potential of fast crops like salad mix. The biology can do a lot of work if you let it.

On the other hand, if you're not in a hurry, I've realized that, when it comes to getting rid of the residue from a previous crop, you can either till or let your soil digest it. This is particularly applicable to some flower and longer crops, where there is no hurry to get rid of them at the end of the crop, because there's not enough time to plant anything after it.

If there's no hurry, crops can simply be tarped down to let biology do the rest. I saw this on my visit to Bare Mountain Farm, where they were tarping down a bed of flowers that had gone by at the end of the season without even mowing it. Why go to the trouble if you don't have to? Since they wouldn't use the bed again until the next year, they knew their thriving soil biology would break down the residue of the previous crop with almost no work on their part. See the interview with Bare Mountain Farm on p. 123.

No-Till Makes It Almost Irrelevant How Bad Your Soil Is

A common theme I noticed in the interviews was that farmers were able to grow on very poor soils by mulching heavily and building soil up, and able to grow on sloping land because they don't have to worry about getting a tractor stuck. I got both of these insights from my interview with Mossy Willow Farm (p. 223) when Mikey told me Paul Kaiser of Singing Frogs Farm's (see interview p. 275) advice to him: "When working no-till on a clay soil, farm above the clay." And then Mikey told me how there aren't many vegetable growers in their area; they're surrounded by vineyards. I realized that no-till was allowing Mossy Willow to farm on clay on a hillside, not normally prime agricultural land.

It's very important to be able to work on less-than-perfect agricultural soils, in order to have a decentralized, localized farming system. By building your own soil up on top of the existing poor soil, you should be able to farm almost anywhere.

Credit: Andrew Mefferd

Vegetable rows were interplanted with flowering plants for farmscaping, like this dill, to attract beneficial insects, like these margined leatherwings.

Skipping Tillage Makes It Easier to Increase the Amount of OM in Soil

Since tillage burns up OM, simply skipping it will make it easier to build soil OM. In addition to sequestering carbon, higher levels of soil OM have a long list of benefits, including promoting soil life and nutrient cycling and increasing the infiltration and water-holding capacity of soils. Higher OM soils are more resistant to extremes of moisture—they hold more water during a drought, absorb water more quickly after rain, and are less prone to washing away in a heavy rain than plowed soils. There are a lot of reasons to want increased organic matter if your soils are low.

Reducing Tillage Should Also Reduce Weeding

Though some growers interviewed claimed more of a benefit from this than others, most of them saw reduced weed pressure over time the longer ground went un-tilled. The less they stirred up the weed seed bank in the ground, the fewer weeds came up, though of course there are always weeds that blow or are tracked in....

Gets You on the Ground More Quickly in Spring

A number of interviewees told me about being able to get on their fields in spring before their neighbors, or even farm all winter long in milder areas since they didn't have to get a tractor on the field for cultivation. This is a big advantage when it comes to early crops, keeping employees through the winter, and having a diverse array of vegetables and flowers for much of the year.

Environmental Benefits

There are a number of environmental benefits that stem from adopting organic no-till growing practices, including reducing the amount of pollution from farm machinery, reducing off-gassing CO_2 and erosion from tillage, and increasing carbon sequestration.

Reduced Necessity for Mechanization

The fact that most of these systems aren't dependent on having a tractor or other heavy machinery will make farming more enjoyable if you don't like driving, fixing, fueling, or hearing equipment.

That said, if you love your tractor AND no-till methods, you could use the roller-crimper method or use tractors to scale up one of the other no-till methods.

Disadvantages of No-Till

Mechanical Cultivation No Longer an Option

You don't have the option to erase weeds with a tiller anymore. If they get out of control, you can't mechanically cultivate anymore. You have to stay on top of weeds and keep them from going to seed or they will

Credit: Andrew Mefferd

Rows of broccoli interspersed with various farmscaping treatments in order to measure the amount of beneficials attracted by different treatments.

get out of control with no good means for getting them back under control. This just has to be taken into consideration when planning for no-till.

Mulched Soil is Slower to Warm in the Spring

Any method that has light-colored mulch on the soil in the spring will warm more slowly than bare soil. In fact, some growers using the deep straw mulch method pull the mulch back in the spring to warm the soil before planting. Just keep in mind that soil-cooling mulches are not the method to use for extra-early crops.

On the other hand, this can be an advantage in very hot areas where the soil could benefit from being kept cool. Though not warm in the springtime, mulched land stays warmer later into the fall according to some growers, so under some circumstances it may be a better technique for the end of the season than the beginning.

High OM Can Lead to Slugs

Slugs and snails can become a problem when there is a lot of undigested organic matter in the soil. Because dead plant matter is what they eat, they will come for the decaying organic matter, and stay for your crop. How to deal with them other than Sluggo? Ideally leave beds uncovered for a few days after occultation before transplanting into, so they retreat. Slugs may be particularly pronounced when you first establish no-till, but the longer you leave the ground undisturbed, their numbers may be reduced over time by ground-dwelling animals like beetles and snakes.

3

An Overview of Organic
No-Till Techniques

How un-tilled is no-till? What exactly is no-till, and does it matter? Many of the growers I met with have their own definitions.

Any reduction in tillage is headed in the right direction. One thing all the growers I talked with agreed upon is that not inverting the soil layers was an important part of no-till.

Important Differences Between
Organic vs. Conventional No-Till

If there is any skepticism about the scale on which no-till can be useful, one place where the transformative potential for no-till is on display is in conventional row cropping. For proof of this look to the two most widely planted crops in the US: corn and soybeans. Consider that over the last forty years no-till has gone from nonexistent to making up nearly half of the acreage of these two major conventional crops.

There is one very big difference between conventional row-crop no-till and the organic methods detailed in this book: the conventional methods depend on both herbicides and genetically modified crops, and will never be available to organic vegetable and flower growers.

As in other areas of agriculture, row-crop farmers have traditionally relied on tillage to remove the residue from one crop, prepare the soil, and suppress weeds for the next one. Traditionally, the area of land you could plant to corn or soy was limited by how much ground you could plow or otherwise prepare before planting time.

The trend in much of corn and soy has been to skip plowing altogether. There are three technologies that have made this possible: no-till planters (often called no-till drills), herbicides, and crops that have been genetically modified to survive the herbicides. No-till planters have made it possible to plant into a rough field that has not been loosened and smoothed by cultivation, and still has residue from the previous crop in it. So the drill gets around the problem of not having the slate wiped clean from the previous crop.

There's still the problem of weeds to deal with, which is where the herbicides and genetic modification come in. Before the advent of herbicides, dealing with weeds mechanically was also a limiting factor in the amount of land that could be planted.

Herbicides of course made it possible to kill weeds by spraying them, but then there is the problem that they will kill the crop, too. The solution that has come to dominate conventional row-crop production is using a genetically modified crop that can survive the herbicides that kill the weeds. With 88 percent of corn and 93 percent of soy genetically modified, it's hard to imagine a more complete takeover of the most widely grown crops.

This works because when the crop and the weeds germinate, herbicide is sprayed over top of both, leaving only the crop standing. I learned about this firsthand growing vegetables in Pennsylvania. Our farm was surrounded by conventional dairy. I saw my neighbor go out and no-till drill his soybeans in the spring. A few weeks later the soybeans were up—and more weeds than beans. A couple weeks later, the weeds had overgrown the beans, and the soy was barely visible through the weeds. In my naiveté, I thought, "he's going to lose his soybean crop to weeds."

Silly me. Shortly thereafter, he came through and sprayed herbicide, and the weeds died and the soy survived due to genetic modification. Now, I think that this solution is problematic for a lot of reasons, and I don't think that vegetable and flower growers should aspire for the same type of solution. But it is proof of the concept that no-till can be adopted very quickly over a wide area when it is advantageous.

It is unfortunate that in conventional no-till, the increased efficiency comes along with an increase in herbicide usage, since the system only works if you are able to spray herbicides everywhere, including all over your crop. Estimates are that glyphosate production increased ten-fold during the period when genetically modified crops and conventional no-till were being rapidly adopted—from 15 million pounds in 1996 to 159 million in 2012.[14]

The other effect of the increase in herbicides has been the development of herbicide-resistant "superweeds." Just as overuse of antibiotics has produced drug-resistant bacteria, relying on herbicides so heavily has bred weeds that survive them. Unfortunately something that is ostensibly good (increasing efficiency, reducing the need for tillage) on the conventional side comes with all the benefits of an arms race and drug addiction at the same time: as weeds become more resistant, farmers have to use more and stronger chemicals to get the same effect. Ultimately, only chemical companies benefit, with revenues from GM seed having increased sevenfold over the same period. And the real prize for the chemical companies—chemical sales—have increased even more.

Credit: Andrew Mefferd

No-till broccoli. You can see how weeds start to sprout as the crimped residue breaks down.

How we could think that dousing the majority of our farmland in hundreds of millions of pounds of any chemical could come without negative consequences is beyond me. This is also why the largest seed companies in the world—like Dow and Monsanto—are chemical companies. They want to sell seeds, but what they really want to sell are the chemicals that go with their seeds. That's why they breed plants that go with their own proprietary chemicals; so buying their seed locks farmers into buying their chemicals, too.

So what ties all the organic no-till techniques together? The answer is the use of mulch; all the different systems use mulch in one form or another instead of tillage. And they all include a step for either killing the weeds in the top part of the soil (solarization and occultation) or suppressing the weeds from germinating (mulches applied or grown in place).

To make sense of them, I've broken the systems into two broad groups based on whether they use biodegradable mulches or not. This is because the type of mulch affects how it is managed. In no-till non-biodegradable (usually plastic) mulches are usually removed before planting the crop, whereas the biodegradable mulches are typically left in place during crop growth.

Each of the two groups can be further broken into two subgroups, for a total of four. The non-biodegradable mulches are broken into opaque vs. clear plastic mulches. Use of opaque mulch is called occultation, and use of clear plastic mulch is called solarization.

The biodegradable mulches can be broken down into whether they are grown in place or brought in from elsewhere and applied to the soil. These methods are not used to the exclusion of each other. Many growers use more than one in conjunction with another one, based on field conditions and what they are trying to accomplish.

Biodegradable Mulch Grown in Place

This is the system that I first learned, where a cover crop is grown and then killed in place, to form a physical barrier between weeds and the crop. It's the same idea as using plastic mulch to suppress weeds around a transplanted crop. Except instead of using a plastic mulch to suppress weeds, the mulch is grown in place and killed before the crop is planted.

I don't focus on this method in this book, because it has been covered by one of the originators, the Rodale Institute's Jeff Moyer in his book *Organic No-Till Farming*. Like the other methods I do describe, it's a very simple concept. The basic method is to grow a cover crop until it is in the flowering stage, and then kill it using a roller-crimper.

This practice is much simpler in conventional agriculture, where herbicides can be used to kill a cover crop at any stage. It is not feasible to let a cover crop grow until its natural death at maturity, because it also sets seed when mature and would re-seed itself.

"As with all things in life, timing is everything. This couldn't be truer when it comes to organic no-till," said Moyer in *Organic No-Till Farming*. Getting the timing right is crucial for this method to work. Planning must go backwards from the planting window of the cash crop. A cover crop must be planted that can be killed when the cash crop needs to be planted.

The cover crop must be planted thickly and have enough biomass to get good suppression of weeds that want to germinate. Planting a

Fennel with margined leatherwings in front of no-till broccoli.

cover crop either requires tillage or having a no-till drill to plant into an un-tilled field. So those wishing to practice this method with no tillage at all need to get or have access to a no-till drill. Rye has become a popular cover crop for this method, because overwintered rye can be terminated at the right time for a lot of spring crops.

When the right stage is reached, the most common method of organically terminating a living cover crop is to use a roller-crimper. A roller-crimper is an implement that consists of a cylindrical drum on its side that is driven over the cover crop to mash it down and flatten it. Just knocking most cover crops over would not

Credit: Andrew Mefferd

be enough to kill them, however, so fins are added usually every 5–7 inches to the length of the drum to crimp the cover crop stems and stop the flow of juices up and down the plant. It is mounted most commonly on a riding tractor, but there are smaller models for use with walking tractors.

Once the cover crop is terminated and flattened, the cash crop can be transplanted into it. Since there isn't a loose, friable seedbed as with a tilled bed, transplants have to either be put in by hand with trowels or by mechanical means.

One of the simplest ways to mechanically prepare a killed mulch bed is to run a tractor over it with a no-till planter. Coulters can be added to mechanical transplanters to cut a path through the mulch for the transplanter shoes. A less mechanical, cheaper method is to use a toolbar with fertilizer knives or something similar attached (not for applying fertilizer, just for loosening the soil), and a straight coulter in front to cut through the residue in front of the knife.

The least mechanical option is to use a trowel to simply dig transplanting holes out of the mulched bed. Shawn Jadrnicek developed a method for using a bulb transplanter to dig transplanting holes out of the bed. This way the holes can be prepared without bending over, which may speed things up (see his guest chapter for more info on this).

When done properly, this method can work very well, with the flattened cover crop preventing weeds from germinating around the cash crop. It is important to get a dense, even stand of the cover crop, for a weak or patchy stand will not suppress weeds adequately. The cover crop needs to be terminated at the exact right time. Done too early and the cover crop may regrow and not form a mulch. Done too late and the cover crop will have mature seed, which will germinate and compete with the cash crop.

Timing is also important when planting the cash crop. It needs to be planted as close to when the cover crop is crimped as possible. The crimped cover crop will only suppress weeds for a finite amount of time, after which it will start breaking down and allowing weed growth. When timed right, some cash crops, like pumpkins or winter squash, may form their own canopy, continuing to suppress some weeds after the mulch starts to break down. Other crops simply get a head start and become established without competition before weeds start to grow.

Advantages of Mulches Grown in Place

This method is advantageous when trying to do no-till on a larger scale. Instead of buying or making a mulch material, all you buy is the cover crop seed and the equipment to terminate it with. In my experience this works well for transplanted vegetable crops that develop into a fairly large plant, like tomatoes and winter squash. For farms looking to grow a larger field of a transplanted crop, like a pumpkin patch, this could be a good technique.

Disadvantages of Mulches Grown in Place

As strong as this method is for larger acreages, it has a number of disadvantages for smaller acreages in mixed vegetable production. In addition to requiring a higher level of mechanization, the mulch keeps the soil cooler than bare soil would.

Credit: Andrew Mefferd

Harvesting broccoli for no-till yield trials. Different treatments were compared to gauge the effectiveness of the treatments.

There was a story in the September 2001 issue of *Growing for Market*, "The Search for Organic No-Till," in which Ron Morse was extensively quoted. Regarding the planting delay, he said, "Not all vegetable crops are suitable for no-till. Planting will be several weeks later than conventional planting dates for three reasons: First, the cover crop tends to keep the soil moist and cold in spring; second, the cover crop has to get big enough to make a good mulch; third, the farther into blooming the cover crop is when mowed or rolled, the more easily it will be killed. Trying to kill a cover crop before it's flowering increases the chance that it will regrow; in fact, some growers wait till vetch is 50 percent in bloom before mowing it."

"For anything where you don't need earliness, you need to take a serious look at no-till," Morse said. "Pumpkins are the classic example of that. No-till is the way pumpkins are grown now in many states."

"Other crops that have succeeded with no-till include cabbage, tomatoes, cucumbers, potatoes, winter squash, and fall broccoli (which Morse likes to grow on a crop of German millet and soybeans)," the article said.

One of the most notable disadvantages is that this method requires more equipment than the other no-till methods. Below a certain farm size, it just doesn't make sense to invest in a tractor and all the associated equipment. To get started with this method, you have to either till a field or have a no-till drill to plant your desired mulch cover crop into an un-tilled field. So the first option of plowing a field to establish the cover crop is not going to work for those who want to be strictly no-till.

The investment in equipment is considerably higher for this method than the other no-till methods. An alternative way to practice the roller-crimper method without owning a tractor would be to do it with a walk-behind tractor, a rototiller, and a roller-crimper. This would require less of an investment than with a tractor. It would probably necessitate tilling to get the cover crop planted, though, because as of this writing there aren't any no-till drills for walk-behind tractors that I'm aware of. So still on a very small scale, the mulch grown in place method would necessitate more of an investment in equipment than the other organic no-till methods.

Along with the higher investment in equipment for this method comes a higher level of complexity and planning. Remember, the cover crop only has a certain window when it can be killed. Roll it too early, and it will re-sprout and spring back up. Roll it too late and it will have set viable seed, and your cover crop will be a weed. So you have to make absolutely sure that rolling happens at the right time or it won't work. Look at Shawn Jadrnicek's guest article and see the great deal of planning that goes into his roller-crimper system. The planning and execution of this method is much more complicated than, say, occultation, where you put a tarp on the bed you want to plant until it's time to plant it.

The higher level of complexity corresponds to a decline in flexibility. The cover crop only has a certain window when it can be rolled, and the mulch only suppresses weeds for a limited amount of time before weeds start growing through. This translates to a limited planting window for crops. This works well for single plantings of cash crops that fall in the planting window for a given cover crop, for example, planting grain or winter squash or tomatoes following a rolled rye cover crop.

However, most small, diversified vegetable and flower farms rely on multiple succession plantings throughout the year to provide a steady supply of a variety of crops. Rolled cover crops are not conducive to numerous small plantings of different crops through the year and quick bed turnover.

Because the soil is not thoroughly pulverized and tilled, as with a rototiller, the rolled cover crop method tends to result in a rougher seedbed than traditional tillage or the other no-till methods in this book. This is not a problem for transplanted crops or relatively large-seeded crops like grains or winter squash, but it can be problematic for small-seeded directly sown crops like salad mix.

Mulches grown in place make the most sense for long-season crops that have a planting window in the spring that corresponds to the weed suppression window of a rolled cover crop. For less upfront investment and mechanization, and increased simplicity and flexibility, see the other methods described in this book. Rolled cover crops can integrate well into a diverse market farm, for example, when planting a pumpkin patch or a field of tomatoes. A crimped cover crop can also be a great solution to provide effective no-till weed control on larger acreages. For most small diversified farms, however, one of the other methods in this book is probably a better solution.

Harvesting an assortment of no-till brassicas.

Credit: Andrew Mefferd

Applied Biodegradable Mulch

Another biodegradable option that is more flexible than mulch grown in place is biodegradable mulch that is applied when and where needed.

Almost anything that biodegrades and stays in place can be used as an applied biodegradable mulch. This can include thick layers of compost, cardboard, paper mulch, unrolled hay bales, straw, or leaves. In many areas, there may be by-products of other industries that could be used for mulches. For example, nut shells, wood chips, spent brewers grains, and many other organic materials might be available locally. If these materials are by-products of other industries it may mean that they are available for free or for low cost. Though the price may be right, it is important to consider the impact on soil health of anything used.

Wood chips, for example, can have great long-term benefits for raising OM and fungal components of soils. However, a large addition of raw wood chips can have a near-term negative impact on growing conditions by tying up nitrogen as the soil biology digests the wood chips.

Almost any organic matter can be put to good use around the farm, as long as it is used properly. In the case of the wood chips, as with much other raw, high-carbon organic matter, you would likely be better off composting the wood chips by themselves, or adding them as the high-carbon component of a compost pile, rather than adding them directly to the soil. (See Tobacco Road Farm's recipe for high-carbon compost on p. 307 and a similar recipe in use by Natick Community Farm on p. 241.)

Advantages of Applying Biodegradable Mulches

Simplicity is one of the advantages of applied biodegradable mulch. In the most basic form, mulch is applied to the growing area before planting. The crop is planted into the mulch and it prevents weeds from growing. This should result in less weeding over time as new weed seeds are not stirred up through tillage.

The greatest number of no-tillers that I was able to get ahold of were somewhere in the applied biodegradable mulch group, aided by solarization and/or occultation. My guess is that this method is the most popular because it is relatively simple, and because most farms already have access to mulch and tarps, so it's not hard to try. Even though this is by far the biggest category, there is a big range of practices within it, from growers who use deep mulches almost exclusively with little or no solarization or occultation, to growers who use tarps predominantly with very little mulch. Look to the interviews for inspiration on how to find the right mix for you.

I'd like to break the biodegradable mulchers down into three subgroups. There is one set of techniques associated with deep straw mulch, a separate method for using compost and other loose mulches, and one for using cardboard.

Mulches conserve moisture by blocking evaporation. They may decrease the amount of watering necessary, so they can be a good option for dry areas. Since, unlike plastic mulches, they stay in place after the crop is removed, biodegradable mulches contribute to building soil organic matter as they break down in the soil.

Disadvantages of Applying Biodegradable Mulches

The disadvantages of many applied biodegradable mulches are the opposite of the disadvantages of biodegradable mulches grown in place. They are usually more time-consuming to apply on a square foot (area) basis, so they may be more effective on smaller areas.

Mulches grown in place take advantage of mechanizing the process of applying the mulch for efficiency, in this case planting and then rolling and crimping the cover quickly by tractor. This is where the scale becomes important in determining the best method. Applying cardboard mulch on an acre, for example, would take much longer than rolling a cover crop using a tractor on an acre.

There are materials and methods that can be used to more rapidly apply a biodegradable mulch, to make this method more time efficient on a larger area. Materials like rolled paper mulch or round bales of hay or straw can be deployed more quickly over a larger area.

Another way to speed the application if you want to use loose biodegradable mulch like compost is to mechanically apply it with a spreader of some type. Manure spreaders typically don't apply an even or heavy enough layer to be useful for this. Making multiple passes with a traditional manure spreader, and raking or otherwise evening out the results, is one possibility. Another is to get a drop spreader, which lays down a much more even layer than a regular manure spreader. See the Resources section for suppliers.

Once the planting area is prepared, for example by using solarization or occultation to kill the weeds, paper or hay/straw can be rolled out over the beds to continue suppressing weeds during the growing season. Seeds or seedlings can be planted right into the mulch.

Non-Biodegradable Mulches

Solarization and occultation are two closely related methods that make use of non-biodegradable mulches to kill and break down the existing vegetation. Occultation is the word for tarping—putting an opaque barrier down to kill vegetation and compost it in place. Solarization is the word for using a clear tarp to cook the weeds.

Occultation

Occultation is stale seedbedding without the flames. If you've ever set a bucket down on a grassy area, forgotten about it, and come back a few weeks later, you've done occultation on a very small scale. If you forgot about your bucket for long enough, a perfect brown circle where the grass died and vanished smiled back up at you when you picked your bucket back up.

Someone realized that if they did this on a grander scale, they would have a nifty way of getting rid of existing vegetation and preparing the ground for crops. The first time I was exposed to it was in Jean-Martin Fortier's book *The Market Gardener*. In it he talks about how he uses occultation, along with tillage, to get rid of vegetation, and as a placeholder to keep an area weed free until he can plant it.[15]

I feel almost silly explaining how occultation works, it's so simple. Though it's worth understanding why and how it works.

Occultation cuts plants off from light, starving them and causing them to die. When black tarps are used, they heat up in the sun,

Credit: Andrew Mefferd

A little of the broccoli we harvested for yield trials. It was donated to a food bank.

speeding up the process. Once the vegetation is dead, all manner of soil life moves in and does what it does: breaks down the now dead organic matter. In the warm, moist conditions under the tarp, weed seeds also germinate and then die due to lack of sun. So occultation can get rid of the vegetation on the surface of the soil, opening the way for planting crops, and also reduce the weed seed bank at the same time. Since, when you are not tilling, you don't bring new weed seeds up to the surface from below, over time occultation should deplete the weed seed bank.

How long occultation takes depends on the temperature: it works faster in hotter weather. Growers interviewed for this book reported anywhere from three to six weeks, depending on the season. Figure out what works for you under your conditions. But keep in mind that occultation does not have to proceed to the point of bare soil to be effective. When they didn't have time for the biomass to break down completely, some growers raked the partially decomposed organic matter off of beds to reveal a plantable surface.

Solarization

Solarization takes advantage of the greenhouse effect to kill whatever is under it. It's as simple to do as occultation: just lay a piece of clear plastic down on whatever you want to kill and leave it there until it is dead.

In the summertime or in a hoophouse, this can be really quick; 24 hours may be all that is needed. Other times of the year it will take longer.

For solarization, all you need is a clear piece of plastic. It's a great use for old greenhouse plastic. For occultation, all you need is an opaque tarp. Used silage tarps can be a cheap, durable, reusable source of tarp material. See the Resources section for more about this, including a source for reused silage tarps.

Occultation and solarization are similar enough that their advantages and disadvantages can be looked at together. They have the same benefits of the other no-till methods, in that they don't need specialized equipment or use any fossil fuels, and are efficient of space. Whether you grow on raised beds or on flat ground, you don't have to remake beds or smooth the ground with these methods, as you would with tillage. You don't have to wait for the ground to dry out to get on the field either.

One of the disadvantages of these methods is that they take some time—more in the case of occultation, less in the case of solarization. In any event, they're not immediate, like say rototilling a bed and coming back and planting it, so they take some planning.

Another disadvantage is that they work more slowly at colder, less sunny times of the year. Even in the summer an unexpectedly cloudy patch may slow the methods down, especially solarization. And depending on your weather, solarization may not work in the winter where you are. Perennial weeds may continue to be troublesome, as deeply rooted ones may have enough reserves to survive the darkness of occultation or the heat of solarization.

Ultimately, occultation and solarization can be great methods for breaking land in and getting started with no-till as long as you use their disadvantages to your advantage. While occultation in particular is not fast, it is a great, easy method to open new ground. For example, if you

Credit: Andrew Mefferd

A much simpler no-till tractor seeder. A coulter in front cuts the residue, while a shank (not pictured) loosens the soil, and the seeders (in back) drop the seed. It could be adapted to setting transplants by hand by taking the seeders off and transplanting into the furrows left by the coulter and shank.

can look ahead and see that you are going to need additional land to grow on next season, throw tarps down the previous fall, or even the previous summer (the longer the tarps are down the better). When the snow melts the following spring, you should be able to pull the tarp back to a fairly clean planting area. If the black side is up, the tarp may even speed up snowmelt in the spring for earlier planting.

Conclusions

No-till is about figuring out how the core techniques of plastic mulches (occultation and solarization), applied organic mulches, and mulches grown in place apply to the individual farm's weather, weeds, and resources. I don't think there's any one-size-fits-all formula; it is my hope that growers can read this book and pick and choose the techniques that will make no-till work for them.

Among the growers surveyed, there was quite a bit of diversity. Some emphasized fast rotations of quick crops, some took their time with longer crops. Some used a lot of compost, some didn't. Some cover cropped, some didn't. Some do everything with hand tools, some had tractors and machinery. The point is they all have found ways to be profitable and build soil, even though their no-till systems may be quite different.

It might seem contradictory to hear both from people who like and don't like raised beds in the same book. But what works well for one grower may not work well for another. Success in farming is such a mix of climate, personality, soils, goals, personal style, and a thousand other things that go into a system, that I wanted to profile what is working and let you decide what to try on the path to developing your own successful system. Here are some tips and themes that developed over the course of the interviews.

Tips and Themes

It's all about the mulches and the biology. All of the no-till systems use some sort of mulch at some point to either smother or suppress weeds. One thing that is going to happen if you are increasing organic matter and not using chemicals is that you will get increased biological activity in your soil.

Increasing biological activity is important because it breaks organic matter down and makes nutrients available to plants. Biology is the driver behind soil function. The more diverse and redundant, the more stable the soil system.

Before starting a no-till system, a lot of growers do something to kill weeds preemptively like mulching with cardboard, straw, occultation, solarization, or even one last till to set the stage for lower weed pressure. The longer weed suppression can go on before starting the no-till system, the cleaner the no-till beds will be. Starting the previous season is the ideal situation. Prepping beds in the fall and tarping them as a placeholder over the winter may be the fastest way to get back in the ground in spring.

Tarps are very versatile. In addition to being used as placeholders, they can be used to hold moisture, and impermeable tarps can exclude moisture during wet times of the year.

Fast bed turnovers will make the most of quick crops like salad mix. Season-long crops like many of the flowers may benefit from the efficiency of simply putting a tarp on a crop in a space that can't be cropped again that season. When going quickly between short crops, if the previous crop was fertilized it may not be necessary to add more fertilizer or compost.

Some growers that use a lot of compost for mulch make or source a high-carbon compost so they can add high amounts of organic matter without excessive amounts of fertility. For recipes, see Tobacco Road Farm p. 305 and Natick Community Farm p. 237. Many growers added a lot of compost in the beginning to raise their SOM quickly, then tapered off compost application or started using high-carbon compost.

Perennial weeds can be a problem for no-till systems because they store energy in their roots, so they can't be smothered as easily by mulch.

The pattern I saw on many farms is that the longer they've done no-till, the better it works. Since much of the weed pressure on farms comes from the annual seed bank in the soil, the longer they got away from stirring those seeds up, the fewer weeds they had.

One way to break in some no-till beds and get a crop at the same time is to grow a vining crop like winter squash, melons, or pumpkins, planted through a tarp or other heavy mulch. The plants can vine out over the mulch while occultation is taking place.

The deep straw mulch and roller-crimper method can work well for hot areas where it is advantageous to keep the soil cool and mulched.

Many of the growers made use of free sources of mulch: wood chips, manure, straw, even beer filters in Urbavore Farm's case. See what's available in your area.

Other Resources

Think about all the plants, animals, and people on your farm. The buildings and infrastructure that give it its feel. One of the most difficult things about writing this book was to capture the feel of being on these farms without making it longer than it needs to be.

Visiting with each one of these farms was incredibly inspiring in its own way. I'm sure every one of them thinks there is a lot more I could have said about them, and that would be true. Many of these farms have their own resources, ranging from online courses, to on-farm workshops, to books. If you are intrigued by any of these farms, there is much more to each of them. I encourage you to seek them out in the Resources section if you want to know more.

To All the No-Tillers Out There

I know I didn't get to you all! Between time and number of pages I knew there was no way to get to everyone. If you are interested in having your no-till story told either through *Growing for Market* or perhaps a future edition of this book, get ahold of me and let's do an interview or a visit.

Grower Interviews

UNFORTUNATELY FOR ME I HADN'T YET HEARD OF Astarte Farm when I was swinging through Massachusetts doing farm visits, or I would have tried to stop by for an interview during the growing season. It wasn't until it was snowing that someone asked me if I had visited Astarte on my quest to interview no-tillers.

So I got ahold of Dan Pratt, and he was kind enough to talk to me about his experience with no-till over the phone in January while the farm was under a blanket of snow.

"I no longer own the farm. I sold it three years ago, but I was lucky to be quickly hired as the farm manager. The whole property is 6.6 acres, but with a house and three barns, and the mandatory buffer zone for the organic certification, we're growing on about three and a half acres," said Dan. "We have a heated propagation greenhouse, 26' by 48'. We have three high tunnels of varying sizes. We have about 85 production beds that we rotate our crops through. The beds are 175' × 3' with 3' pathways between beds. The pathways are either mown weeds or cardboard with wood chip mulch."

Dan, who was already an experienced farmer, bought the farm in 1999 and sold it in 2013. Ever since, he has been working to encourage long-term fertility by using biochar, pollinator and predator habitats, and no-till.

"How did you get interested in no-till in the first place—you started out doing more of a conventional tilling system, correct?" I asked.

"Yes, in a sense. I started out with a little Italian spader, a 36-inch Tortella model. The rumor was that spading was much easier on your soil structure, and might not completely obliterate your earthworm population, because it wasn't churning the soil and inverting the soil profile. I started doing semi-permanent production beds, where we would run the tiller, and then we had grass or weed paths that were undisturbed," said Dan.

ASTARTE FARM

Dan Pratt
Hadley, Massachusetts
Mixed vegetables
*Occultation, mulch grown in place,
and applied organic mulches*

I know the spader well, because when I realized that roller-crimper no-till didn't scale down to my original farm, I bought one in an attempt to avoid some of the problems with repeated rototilling. It's called a spader because it has a row of miniature shovels that dig into the ground and throw a chunk of soil against a back plate to pulverize it by impact, instead of by churning the soil under like a rototiller.

"It was roughly a three-foot path, and a three-foot tilled bed, and a three-foot path, and on like that, thinking that at some point, I would let everything switch [from beds to paths and vice versa] over so I'd be able to get into that fallow ground in five years. But the fact of the matter was that after five years of growing and mowing weeds, I had a weed seed bank in those paths to beat the band," said Dan.

"So whenever I would turn one of those paths, I faced really tremendous weed pressure. The spader was a pretty good little tool, but what I noticed after about six or seven years of doing this was that my grass paths, which had just been mown, were rising, and even though the production bed would look like chocolate cake after you'd run the spader through there—fluffy, wonderful soil, easy to transplant into, great for direct seeding—after a couple of good, hard rains, the production beds were sinking two, three, and four inches below what had been in sod.

"And then I heard Elaine Ingham at one of the NOFA summer conferences. This really caught my attention because of her talk about soil organic matter, and having everything you needed in the soil. We're lucky to be in this Connecticut River Valley. We're farming the bottom of an old lakebed. There used to be an ice plug in the Holyoke Range that held back a lake for ten thousand years. This soil is probably some of the best soil in the world.

"I mean, if you find a rock out in our field, you know somebody carried it out there and dropped it. It's super easy to work with, but here was this evidence in front of my eyes. I don't really do a lot

Getting ready to roll down some winter rye with a front tractor mounted roller-crimper.

Credit: Dan Pratt

of soil testing, but I do it every two or three years. I was just staying flat on my organic matter, even though I was adding compost and cover cropping in the winter. I was just going in there with that little spader and tearing up everything that was trying to get established in the soil. I foolishly thought that these grass paths would be like refuge strips, a place for the earthworms to go scurry off into when the spader's coming," said Dan.

Getting Started with No-Till

"It was a bit of a leap of faith to just say, 'Well, let's see what we can do with this. Let's give it a shot.' We started out with a fall garlic planting. We wanted to use buckwheat as the smother crop, but it was one of those falls when you could not get buckwheat to die, because we didn't have a killing frost in the fall. We ended up mowing it multiple times with a little mulching mower, which was laborious, and then laid down a couple inches of compost. After that we used Weed Guard Plus paper mulch," said Dan.

"The paper was pre-punched at seven and a half inches in three rows, with a total 36-inch width, so we just laid that down right on top of that two inches of compost and held down the edges with aged wood chips. Then we poked our garlic cloves through the holes, laid about an inch of compost on top of the paper to keep it from blowing away, and we were off and running."

"So you didn't even try to get the cloves down into the soil? You just poked them into the compost, you didn't try to get them into the soil underneath?" I asked.

"No, we just pushed them in as far as they would go through the compost and mowed buckwheat, that was as far as we went with it. I'd like to emphasize using the wood chips to hold down the paper edges as we rolled it out. Because we're in a windy location, we have to be really careful with that paper. It would like to travel two states if we gave it a chance in a breeze," said Dan.

"We had a great garlic crop that came off those first no-till plantings. That was interesting to see as we were just still getting started

I just hate to see us continuing to do this much violence to our soils. It sounds almost sentimental or something, but when you've treated a piece of land right for a little bit, you begin to see so much more life.

— DAN PRATT

with no-till. We hadn't fully committed the whole farm to it, but when we pulled that crop around July 15th, we were able to replant lettuce transplants right into that bed because it was virtually weed free.

"We had a banging crop of lettuce, and just went right ahead and planted lettuce again, and we had a very good second crop of lettuce off those same six beds. That kind of sparked something in us. There was, I would say 10 percent of the weeding that we've ever done before on any production bed at the end of that full year. So it was the second crop that needed the most hand weeding. The first one basically just went in and we were golden. That's how we got started," said Dan.

"Was it weed seeds that blew into that second crop, or was it perennials that eventually came up through the layers of compost and paper that you had put down?" I asked.

"Well, when you pull your garlic out of the ground, you're always pulling some dirt up on those long roots. We always try to shake them as we pull them up, try to keep all the dirt in place, but some always comes up. Then, when you have a crew out there transplanting, there's always somebody that flicks some dirt out on top of the compost. We're

Beds after occultation in the foreground, and during in the background.

Credit: Dan Pratt

very well surrounded by conventional growers, and a lot of times when the sweet corn's been picked, well, that's it for that field until they put their fall cover on, so there's plenty of weed seed blowing around in the valley as well," said Dan.

Killing Cover Crops with the Roller-Crimper

"Pretty quickly, we ended up with a very small roller-crimper [based] on that Rodale model. We had it mounted on a quick-hitch plate where the loader would be on our little Kubota tractor. Which was pretty sweet, because then you could lift the front wheels off [the ground], to maximize downward pressure and avoid having to fill the drum with water. We use the individual [wheel] brakes to steer. The thing that we ran into, and it was probably the second year when we were doing no-till, was the root mass that we had left over from the winter rye was just about impenetrable," said Dan.

"We had really nice, sharp soil knives to cut our little holes for transplanting into, and after the first half-dozen beds, the crew said, 'If you ever plant winter rye again, you're looking for some different people to transplant because that's ridiculously hard.' And it really was.

"We should have known that from our experience tilling in winter rye with the little spader. If you let that rye get up over knee high, you were looking at a real job and a half. Because it would just break up into clumps, and the clumps, all or half of them, started growing again, you'd be spading two and three times if you let that rye get very tall.

"What we have used the roller-crimper for with some success is on our cover crop cocktails where we've got oats, peas, and radish together, for instance. We can seed that in the spring and grow it for mid-season crops and have pretty good luck planting right into that. A lot of times, we are putting in a four-inch pot. I have to be clear about that. We don't do that very often for our smaller lettuce transplants or anything," said Dan.

"So you're planting larger transplants like tomatoes or winter squash into crimped oats, peas, and radishes?" I asked.

"Yes," said Dan.

"I'm kind of surprised that the roller-crimping is killing the radish. It sounds like it does?" I asked.

"Well, mostly the radish we're using is in the late summer and fall, so in the early spring, we're just doing oats and peas. And it is killing them pretty well. I did kill some radish, one season when it was so wicked dry that we couldn't get oats to germinate, so I got ahold of some teff seed, and we were using teff and radishes. This is prior to a garlic crop. The teff, I swear, that germinates in dust. I don't know how it does it, but it does it. We had a lovely teff stand, which was frost-sensitive, but in our enthusiasm to do this, we had all those daikon radish out there that didn't kill," said Dan.

"But when the radishes had pushed out of the ground three to four inches, and then you run the crimper over them, it pretty well breaks them all off. I think I probably had to go back down the row and whack a few with the side of a shovel that didn't break off completely, but it was fairly effective," said Dan.

"Okay. Is that a pretty regular thing for you now? It sounds like you're using the roller-crimper for transplants only?" I asked.

"Yes."

"That was my impression. What I came away with after working with that style is that it was great for transplants, but you're not going to direct seed salad mix into a bed that you roller-crimped," I said.

"Right, I'd say that's correct," said Dan.

Timing Is Everything with Crimping

"Really the only problem that we had with those spring cover crops that we put in is timing. We had two lettuce beds that were going in that had the same cover crop seeding date. One of them got rolled on the correct date, and one of them, for weather or vacations or whatever, didn't get rolled for another ten days. [On the

Adding compost post-occultation with the drop spreader.

Credit: Dan Pratt

bed that didn't get rolled on time], we had a regrowth of the oats that was substantial, and we basically ended up with miniature lettuce heads growing in a forest of oats. We were lucky to have a deli that would take those little heads for their salad bar. We didn't lose all our money on it and that was one bed we didn't have to re-seed [after the cash crop], let me put it that way," said Dan.

"That's one of the other limitations with crimping, that you have to be on top of your timing because you have limited planting windows. If you miss rolling at the right time or planting at the right time, you either get regrowth of the cover crop or you don't get enough weed suppression," I said.

"On my wish list for this year is a flail mower," said Dan.

"So you can just drop a cover crop in place?" I asked.

"Yes. I don't know if it's going to chop things too finely to make an effective mulch. It may require using paper mulch again if that's the case, that it's just chopped too fine to provide any weed suppression, but I would like to try running a four-foot flail mower down some of these really nice, lush cover crop beds and see what kind of a seedbed we get from that," said Dan.

"A lot of these guys, like Bryan [O'Hara] (see interview p. 305) and Ricky [Baruc] (see interview p. 81) have figured out these systems that really work for them, and we are still feeling our way into this.

"I would like to say in all confidence, I know that this is going to work. But after three years, the soil hasn't softened up as much as I expected it to, and we've certainly used enough compost as mulch and other things. We've completely abandoned using all blended organic fertilizers, and we're still getting really good growth and yields. There are pluses and minuses, but I wouldn't say we're the experts in this by any stretch in the imagination."

Occultation and Solarization

"We have some grass paths left, but we got sick of doing all that mowing. There's always that month period where once a week isn't enough, and because we wanted to do as little cultivation as possible, the intrusion

Credit: Dan Pratt

A young cover crop of oats broadcast by hand onto compost spread on a previously occulted bed, then five rows of field peas were planted with an Earthway push seeder.

of the grasses just countered the whole thing we were working on," said Dan.

"We had gone to putting down nice, clean cardboard and using wood chips to hold it down. Our biggest limitation has been finding clean cardboard. By that, I mean because we're certified organic, we can't have any cardboard with colored printing on it."

"Black is okay?" I asked.

"Yes. Black is carbon-based ink, but any kind of color doesn't meet the national organic standards," said Dan.

"We bought a little ABI Elite spreader that's a hydraulic drive spreader, and we could lay down as little as a half inch, or as much as two inches of compost, in just a straight drop pattern."

"And that's a drop spreader?" I asked.

"Yes. ABI is the manufacturer. Their elite model is a manure spreader for small to mid-size stables. It has a hopper that holds about two and a half yards. Because it's hydraulically driven, instead of ground driven, by varying the motor speed and your ground speed on the tractor, you can really adjust it to drop the amount that you want. Because it drops straight down, it's not being spread out all over the paths. It's really been a sweet tool for us," said Dan.

"We used to do all our spreading by hand, just two people walking backwards in front of the tractor with shovels. It was ridiculous, the amount of time they were waiting around in the field while I was getting one more bucket of compost. That spreader turns a three-person operation into a one-person operation."

"So you said you're using occultation, and you're using solarization, too?" I asked.

"We did our first major solarization this last summer. It's not quite a quarter acre, but it's a fairly large patch that cuts across our field that's going to be all perennial flowering shrubs. Perennial flowers specifically for beneficial habitat," said Dan.

"But you're using the landscape fabric for occultation for annual crops?" I asked.

"Yes, that's correct. We probably have ten strips out there right now under the snow that will be the early spring plantings. We rotate them right through the field. We have 85 production beds, we'll rotate those through. Sometimes, we use it on top of a rolled strip if we're not going to get a chance to plant into it quickly enough," said Dan.

"Kind of a placeholder?" I asked.

"Yeah, just as a placeholder. It keeps any weed germination from happening. We really like the porous aspect of the ground cloth strips, the fact that rain goes right through it, oxygen goes through it. We use it, for instance, in the high tunnels after we've harvested the crop instead of pulling the plants. Let's say we had early peas. Instead of pulling the peas, we'll either cut them or just mash them down and roll an occultation strip out on top of that, and it makes a lovely seedbed to plant into. Again, we mostly use transplants when there are a lot of crop residues," said Dan, "but occasionally we will just rake the residues onto the bed shoulders, drop some fresh compost and direct seed into that."

"How long do you have to leave that on before you're going to get that lovely seedbed?" I asked.

"It really depends on the season."

"Shorter in the summertime, longer in the fall?"

"Yeah. We've had them out probably for close to five months over the winter, and in some cases [in warmer weather], we're able to get them up in three weeks. Basically, we just go down and keep peeking under there to see what we got," said Dan.

"When you go from one crop, let's say the peas, and you're going to put a tarp down for a few weeks before planting into something else, you said you got away from the blended organic fertilizers. Are you pulling the tarp off and then putting some compost on before planting the following crop, or how do you handle that?" I asked.

"We've definitely done it both ways. We've applied compost and put a tarp on, or an occultation strip, and we've done it after pulling the strip off. It really depends on what's available in terms of manpower and how much compost we have on hand, et cetera," said Dan.

"The only other thing that I'm thinking that might be of interest is the way we tend to seed our summer covers. If we've had a spring crop, let's say it's lettuce, and we cut the lettuce at the soil level so we've got a flat bed, we'll go in there with the ABI spreader and try to lay down a nice, even two-inch compost layer on top of the bed. Then we'll hand-broadcast oats on top of that compost, and not even attempt to cover it at all.

"But when we put our peas in, we use the old trusty, Earthway [seeder] through there with the peas, and that little tiny shoe throws enough compost on top of the oats to get germination, and we can do three rows of peas down the middle of a three-foot bed and a row on each edge of radish, if we want to have radish in there if it's for a fall planting. It just does a super job."

"That's a good tip."

"Not a no-till planter, but works just as well," said Dan.

"Just by planting into the loose compost on top?" I asked.

"Yes. We've tried to rig a piece of equipment with a couple of old hilling disks on a frame. That was the first year when we had that really heavy winter rye cover, and the crew wasn't happy trying to transplant right into it. We tried setting that up where it would cut two furrows, 18 inches apart through the rye, and then run the Earthway in there and plant large-seeded crops. But I don't know if it was because I didn't have the wavy coulters, or I didn't have enough weight on it, and those roots just wouldn't allow the thing to cut properly. We never really had any luck doing direct seeding that way," said Dan.

"Have you gone completely over to no-till? Are you no-tilling parts and still spading certain parts?" I asked.

"We don't use the spading machine for anything except for a wheel weight when we're plowing snow in the winter," said Dan.

"So is it correct to say that you're handling all your planting needs by either the method you were planting your garlic, with the big compost applications and planting paper, or occultation?" I asked.

"Yes, we're 100 percent no-till now."

"Okay. It's interesting that you made the transition over time," I said. "Is there anything I should have asked you, or anything else you'd like to say about it?"

"There's probably a lot more I'd like to say about it. I just hate to see us continuing to do this much violence to our soils. It sounds almost sentimental or something, but when you've treated a piece of land right for a little bit, you begin to see so much more life. The amount of spon-

Planting tomatoes into a crimped cover crop.

taneous mushroom fruiting that we get on our farm after just three years of this is really very, very encouraging to me," said Dan.

"I think one of the biggest side effects of all the tillage and cultivation we've done is ending up with these bacterially dominated soils. And the mushrooms are where the really neat stuff around the root hairs happens. The more of that we can get going, the happier the whole scene is going to be."

"Are you taking any steps in addition to no-till to try and encourage the fungal aspect of the soil?" I asked.

"We've used powdered mycorrhizal inoculant. We spent a lot of time pre-inoculating some of our biochar when we first got started with that," said Dan.

"I was telling you how great this soil is in Hadley, but after probably twenty years, the twenty years prior to us buying the place, there

Credit: Dan Pratt

were just giant disc harrows run over the field twice a year, once in the spring, once in the fall. Every time it was going down to the same depth. The first crop of carrots that we grew on the farm, they all grew down six inches and took a left-hand turn. They came out like little Js. We had standing water in the field, and this is really some well-drained soil that we've got, but it was just a plow or disc pan that was under there. That pan has now all been broken up by earthworm tunnels, and ground beetles, and some deep-rooted weed crops too, I'm sure. I'm one of those people that likes to see a dandelion because I know what's going on deep down in there.

"And we don't have any standing water on the farm at all. It's just good to see those macro results, if you will, and know that that's a result of a lot of microactivity," said Dan.

I called Shawn Jadrnicek on a busy day between spring rains. I wanted to talk to him because he has made innovations in the roller-crimper style of no-till that are important for growers wanting to use this system.

Shawn is the manager of Wild Hope Farm, and he's done a lot of things over the years—he's been a farmer, extension agent, arborist, and landscaper, among other things. Which may be why he is such an original agricultural thinker: He's seen things from so many different perspectives that he brings a new approach to the age-old agricultural problems. Like how to deal with weeds, for example.

I became a fan of Shawn's because of his excellent book *The Bio-Integrated Farm*. In addition to other agricultural innovations, he talks about how he uses crimped cover crops to suppress weeds (see p. 283–85 of his book for Shawn's description of his no-till method). I was especially excited when he sent me an article for *Growing for Market* magazine about some modifications he made to the crimped cover crop system that helped extend the weed-free period.

Since I already had a description of his system in his own words, he was kind enough to allow me to republish the article as a part of this book. But I had some questions about the method so I wanted to check in with him about what he was doing.

"Would you tell me a little bit about Wild Hope Farm?" I asked.

"It's 212 acres, we basically just started farming it last July. It's in Chester, South Carolina, and it's owned by the Belk family, Tim and Sarah. Right now, we're primarily vegetable production. Our main market is CSA. This year we're putting three and a half acres into vegetables," said Shawn.

"And we have pastured hens on another few acres, that are rotating through some cover crops that we planted. Eventually we're planning on expanding up to about eleven acres of vegetables and adding sheep

Shawn Jadrnicek
Chester, South Carolina
Mixed vegetables
*Mulch grown in place
with added leaf mulch*

A mechanical transplanter with water barrel planting into crimped cereal rye. The mechanical transplanter requires transplants sized at 1.5" and works best with the Speedling transplant trays. With Speedling trays the transplants are pulled out from the top without poking them out the bottom so you can move at a good speed. It's important to have strong transplants that are fully grown and have completely filled out the cells.

Credit: Shawn Jadrnicek

and cows, and pigs to rotate through the fields. This year, our goal is to have a 100-share CSA, and we're hoping to expand to a 400-share CSA over time. We also sell to restaurants, and we're selling at a farmers market as well."

"And are you doing a lot of that no-till, with the techniques you described in the article?" I asked.

"Yes, absolutely. We're actually getting our roller-crimper at the end of this month. So it'll be here just in time for our main crimping operation. When I worked at Clemson University, I had an amazing free resource of leaves. We were inside the city limits, and it was more convenient for the city to drop the leaves at the farm," said Shawn.

"So now, I'm having a harder [time sourcing leaves]. We're just a little too far out of Chester's town limits. So we hired someone with a dump truck to bring the leaves out to the farm. We paid about $5.50 per cubic yard, and paid another $400 for delivery—that's the cheapest way. But we're also planning a logging operation on the farm. We're hoping to get about 1,200 cubic yards of wood chips from that, which we will incorporate into our no-till system.

"I've only used leaves with the system, I've never combined wood chips, so it will be interesting to see how wood chips work out."

"Well, I was really impressed with your no-till system, that you figured out how to take the roller-crimper and improve the length of weed suppression by incorporating leaves as mulch. And remind me, are you chopping those leaves up before you're using them as mulch?" I asked.

"The city has a machine that shreds them as they vacuum them up into the dump truck. So they do come shredded," said Shawn.

"Oh that's handy."

"Yes, it does seem to help, otherwise they blow around too much," said Shawn.

"I interviewed some other people who were using leaves as mulch, and they were pretty adamant about shredding them. How did you hear about the whole roller-crimper thing? I'm always curious where people got their ideas from," I asked.

"It's funny, because I first tried no-till about twenty years ago, way before the roller-crimper was invented. [People] were doing it with undercutters, that was kind of a new innovation, I was doing it with a sickle-bar mower. I would just cut the cover crops with the sickle-bar mower when mature, and then I would rake them into these bundles, and plant my seeds or transplants right next to the bundles. It was kind of time-consuming and I was only farming on a two-acre scale. But then I ended up getting out of farming and becoming an extension agent," said Shawn.

"[And I heard about the roller-crimper method.] I think it was a Carolina Farm Stewardship Association conference, and they had invited Jeff Moyer [now the Executive Director of the Rodale Institute] down to give a special talk just to the extension agents. As I was talking with him about it, I thought, 'This is the best thing that's ever happened to agriculture.' And it really inspired me to get back into farming after seeing what he invented, and what they were doing. I was so excited about that, it just solves all these problems that I experienced.

"I can't remember what year that was, or when he came down to give that talk, but it just blew my mind when I first saw that."

A field a few weeks after mechanical transplanting winter squash, cantaloupe, watermelon, and summer squash into crimped cereal rye.

The same field as above in a later stage.

Credit: Shawn Jadrnicek

Harvesting eggplant from a no-till field with leaves added using a manure spreader to improve weed suppression. Less than two hours of hand weeding was needed in this half acre plot over the entire season.

"So is the article still pretty consistent with what you're doing now?" I said.

"Yeah, everything is still exactly what I'm doing now," said Shawn.

"For the roller-crimper, it has to be a specific scale. I think you're right about that. Because the edges of the crimped areas usually get weedy with a no-till system, you lose about five feet around the edge of wherever you're crimping. So I try to crimp as large an area as possible at a time, making the blocks as wide as possible. But of course if you're adding leaves then it doesn't really matter, because leaves make it perfect everywhere," said Shawn.

"I thought that was a really nifty thing you did with the leaf mulch. Because I know when we were using the roller-crimper system at Virginia Tech, one of the challenges was, if the cover crop wasn't dense enough, then it wouldn't suppress weeds for long enough. Is there any advice you would give to someone trying to get started with roller-crimper no-till?" I asked.

"Definitely, plant your cover crop at the right density and at the right time. And I think the key is really making sure that your cover crop has enough fertility to it. Even if you have the right planting density, but your soil isn't fertile enough, you don't have enough nitrogen mainly, the cover crop is going to be scraggly," said Shawn.

"That happened to me this year, where the fertility was drastically different in the upper half of a field than the lower half. So I got this beautiful, dense, lush green cover crop on one side, and then the upper side is small and scraggly. Normally, that would freak me out because it's going to be difficult to separate the field into two areas, and it's already too late to not no-till. [After a certain point] you're committed to no-till because it's too late to till and prep soil and get ready for planting.

"But I have all of this leaf mulch on hand, so no big deal. I'm just going to slow the tractor down and apply a little more leaves in weak cover crop areas. Short-season crops won't get any leaves when the cover crop is dense and long-season crops get leaves regardless. Adding leaves with the manure spreader allows you to fine-tune your whole system and make it applicable to a wide variety of crops, even with a weak cover crop.

"I would probably only do organic no-till with winter squash, summer squash, tomatoes, and cucumbers if I didn't have the ability to add leaves. Everything else requires a longer growing season, like watermelons, cantaloupe, sweet potatoes, peppers, and eggplant. So really, unless you have the ability to beef up [the roller-crimped mulch] and add some leaves to that, it won't be easy to do [those longer-season crops], because they will get weedy later in the year.

"The [additional leaf mulch] allows us to increase the amount of no-till that we do drastically. I think at Clemson, my last year there, almost 70 percent of our crops were done no-till."

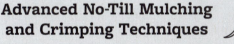

Advanced No-Till Mulching and Crimping Techniques

by Shawn Jadrnicek

Organic no-till and mulching systems are a huge time-saver on our farm. They've allowed us to increase production while reducing labor on our six acres of vegetable production. This spring was our easiest year yet even though an illness in April made it difficult to leave the house. I owe the time savings to the roller-crimper.

The roller-crimper is a heavy metal drum with dull blades that rolls over cover crops, lying them flat and crimping the stems to help kill them. The cover crop remains as a weed-suppressing and water-conserving mulch and transplants

This article originally appeared in the September 2017 issue of *Growing for Market*

Credit: Shawn Jadrnicek

From left, a planting of tomatoes, eggplant, peppers, beans, and sweet potatoes planted into crimped cereal rye. Two inches of leaves were added to the eggplant and pepper bed using a manure spreader to extend the weed control. One inch of leaves were added to the beans and sweet potatoes because they were planted four weeks after crimping and Shawn needed to keep them weed free a little longer but not as long as the eggplant and peppers.

or seeds are planted through the mulch without tillage. However, for crimping to work, the cover crop must be mature, limiting the technique to several weeks during the spring. In addition, weeds are only suppressed for six weeks preventing no-till from working with long-season crops like eggplant and peppers.

I've invented several techniques to make organic no-till more widely adaptable for farms, overcoming some of the hurdles of organic no-till along the way. In the first technique, I plant cover crops on tractor-pulled raised beds and terminate them with a traditional roller-crimper. The raised beds appear to promote early maturity and better cover crop and subsequent cash crop growth in our loamy soils. The second technique delays cereal rye maturity by up to two weeks, extending the planting time frame for no-till. I use a manure spreader in a third technique to add leaf mulch to the crimped cover crop, extending the weed suppression time frame from six weeks to six months. And a fourth technique crimps cereal rye cover crops up to two months earlier by adding leaves on top to prevent regrowth. By combining all of these techniques, I can grow nearly all our crops using organic no-till or mulching systems.

Weed-Free Cover Crops in Raised Beds

The first step to a successful no-till program is growing a weed-free, vigorous cover crop. To do this, I plant all my cover crops into a raised stale seedbed. First, I make the raised beds using a tractor-drawn bed shaper. If fertility is needed, I add the fertilizer prior to making the raised beds or grow a leguminous cover crop prior to making the raised beds. Next, to make a stale seedbed, I wait for at least a half an inch of rainfall or irrigate the field thoroughly. The rain promotes weed seed growth on the raised beds. Once the beds have dried enough to cultivate, I bring the tractor cultivation equipment over the raised beds to kill the weeds. The cultivation toolbar is equipped with sweeps and side wings for the furrows, crescent hoes for the sides of the raised beds, and S-tines for the tops. The idea is to cultivate as shallowly as possible to kill any weeds without bringing up more weed seeds from deeper in the soil. Stale seedbedding once works well, but twice is best.

I perform the last stale seedbedding when the moisture in the soil is just dry enough to perform the cultivation. I then wait between 4–12 hours to ensure the weeds have died from cultivation then plant into moisture. I apply the seeds with a spinner spreader or bag seeder and then re-bed with the bed shaper to incorporate the seeds.

I read about "planting into moisture," but never understood how powerful this technique is when performed after stale seedbedding. Stale seedbedding dries out the surface and leaves the soil loose and unable to give the seed-to-soil contact needed for germination. Cover crop seed planted into the moisture beneath the dry soil surface germinates, while the weed seeds on the surface sit idle from the dry stale seedbed. The cover crop seeds then start growing before the next rain occurs, getting a huge head start on the weed seeds. Once the cover crop starts growing, the plants release weed fighting allelopathic chemicals preventing weed seed germination. The chemicals combined with stale seedbedding create a completely weed-free cover crop.

Cover crop seed selection, planting density, and timing also play critical factors. For winter cover crops terminated in spring, I plant a cereal rye variety called Abruzzi at 150 pounds per acre. I plant the cereal rye six months before our first fall frost date. Planting early gets the cereal rye going before fall cool season weeds start growth. Some of the summer weeds may germinate, but the first frost will kill them before they set seed. Planting earlier or later creates problems with either the warm- or cool-season weed groups. If nitrogen is needed, I've added crimson clover. However, clover breaks down quickly, possibly decreasing weed suppression when used as a mulch. I now add nitrogen to cereal rye before planting by growing a cowpea cover crop. For summer cover crops to be crimped, I use Japanese millet at 30 pounds per acre combined with sunn hemp at 50 pounds per acre and focus on stale seedbedding preferably twice before planting. I crimp the summer cover crop mix according to millet maturity.

Extending Weed Suppression

Since crimped cover crops only suppress weeds for six weeks, they work best with fast-growing crops like winter squash, summer squash, and cucumbers. Longer-season crops like tomatoes, eggplant, peppers, sweet potatoes, cantaloupes, and watermelons grow for such a long period of time that weeds eventually intrude, creating a mess. To solve this problem, I add a layer of leaves to the crimped mulch to extend the weed suppression. The leaves are delivered free from the adjacent city in trucks. The trucks vacuum, shred and compile the leaves into specialized dumping bins holding over ten cubic yards. Once stockpiled, our 125-bushel PTO driven manure spreader then applies two inches of leaves over a quarter acre in two hours. Two inches added to crimped cereal rye gives around six months of weed control depending on cover crop density. Adjust down depending on the crop planted into the mulch, but even one inch will drastically improve weed control and help fill in weak areas in the cover crop.

To extend the weed suppression in no-till with mulch, I first crimp the cover crop, then let it dry out for a few days to become carbonaceous and resistant to decay. Leaves are then applied before or after planting transplants. When applied on top of plants, I unbury the plants by hand after application. Also, some plants such as peppers don't like mulch up against their stems, so I carefully pull it back. Be prepared for mulch to delay plant maturity by about one week.

Mulching with a Manure Spreader

Manure spreaders apply mulch quickly and evenly over the soil, making them the perfect mulching machines. However, our ABI manure spreader is slightly wider than our tractor. When we make raised beds this means that the tires on the manure spreader don't follow the tractor furrows but travel along the edge of the adjacent raised beds. Depending on planting arrangement, raised beds may need wider spacing to accommodate the manure spreader.

Applying leaves on top of crimped cover crops means a small amount of leaves can make a big difference. However, applying leaves on bare soil in larger amounts works well for weed suppression and has many applications. With garlic, I plant three rows per raised bed into bare soil. Just prior to the garlic emerging, I add four inches of leaves to smother the weeds. The garlic pushes through the leaves and the weeds never have a chance.

This saves us around forty hours of weeding in our one-fourth acre of garlic. With potatoes, leaves applied prior to sprouting will prevent weeds and conserve moisture, but also delay potato emergence and maturity. For early emergence and maturity in potatoes, apply leaves after the first cultivation or hilling. Storage onions are the final crop I've used with leaf mulch alone. I apply a layer of mulch to beds prepared in the fall. The mulch is applied to a four-inch depth and remains fallow without a cover crop through the winter. The leaf mulch keeps the weeds at bay until I'm ready to plant the onions in late winter or early spring. I then plant the sets directly into the leaf mulch or just below the soil because both methods seem to work well and the onions stay weed free until harvest in June.

With early-season plantings of summer squash, cucumbers, and cantaloupe, soil warmth is important. Growing early crops on bare soil for the first few weeks encourages earliness from increased warmth. Delaying leaf mulch application until after the first cultivation takes advantage of bare soil warmth while also providing the benefits of mulch.

One year, it was too wet to get into the field and cultivate the crops. However, a break in the weather dried the field out enough to drive the manure spreader over the plants. I applied the leaf mulch with the manure spreader while the weeds were one inch or less, smothering them. My crops were just tall enough to stay above the layer of leaves and pushed through with a little help. The leaf application saved us from a weedy mess the wet spring would have caused.

Using Leaf Mulch to Crimp Early

Many farming inventions and lessons are happy accidents. This one happened when I crimped a cereal rye and crimson clover cover crop about two weeks earlier than I should have. The cover crop wasn't mature enough and popped back up, smothering the tomatoes I'd planted. However, I applied leaves to part of the crimped area—and to my surprise the cereal rye stayed down and died and the crimson clover pushed through the leaves and continued to flourish. I realized with this accidental experiment that applying leaves allows early termination of some cover crops.

To better understand the potential of the technique, I performed two replicated trials using four different mulch depths and two different age classes of mulch on top of cereal rye crimped at three different stages of early maturity. The younger leaves were collected in fall and the older leaves were collected from the previous year's fall. I applied the leaves one, two, four, and six inches deep. My observations indicate that the older leaves worked better at preventing cereal rye regrowth than the younger leaves. I noticed that the wind blew the younger leaves around possibly opening up areas to light. Older leaves congealed together, creating a light-blocking mat. Although the one- and two-inch depth mulch did not prevent cereal rye regrowth, four and six inches of mulch did.

Immature cereal rye is more succulent than mature cereal rye. Because of this, the weight of the crimper must be reduced as much as possible to prevent the crimper blades from cutting through the cereal rye. Once the grass is cut, it easily pokes up through the mulch and continues to grow. I emptied all the water from the crimper to reduce the weight before use. Also, early crimping probably changes nutrient dynamics with decomposition. The succulent nature of the young cereal rye likely promotes a more favorable release of nitrogen for subsequent crops, compared to more mature cereal rye.

I saw little difference in the timing of the early crimped cereal rye. The first crimping occurred Feb 21 when the cereal rye was twenty inches high, just prior to the boot forming (stage 9); the second occurred on March 6, as it was just starting to head (stage 10–10.1); and the third on March 20, when the cereal rye was heading (stage 10.5). The earliest treatment had the most regrowth. However, the crimper weight was too heavy during this treatment and was adjusted for the later treatments.

Shawn using a bulb planter to transplant by hand into crimped cereal rye.

Credit: Shawn Jadrnicek

I also performed a single treatment with two replications on a cover crop of mustard crimped on March 6, just prior to the mustard flowering. With the mustard treatment even the one-inch mulch depth was successful. The control plot that was crimped without adding mulch to the mustard stayed down but weeds quickly invaded.

Planting into Mulched Systems

We recently purchased a mechanical transplanter with a no-till attachment to help facilitate planting into the mulched systems, but I haven't tested it out yet. Previously, transplanting into mulch proved more time-consuming in comparison to bare soil systems, but still saved time in the long run with less tractor work and weeding. Since cereal rye dries the soil out, I pre-irrigate with a drip line if needed to moisten the soil prior to transplanting. I then use trowels or a bulb planter to remove soil in the planting hole for the transplants. The bulb planter is a specialized item from A M Leonard that has the ability to push the plug of soil out without turning the bulb planter upside down to save time. Ideally the bulb planter is the same diameter as the transplant plug.

Planting into the early terminated cereal rye with mulch on top proved more difficult than I expected. When the leaf mulch is moved out of the way to plant into the ground, cereal rye is exposed to sunlight and is difficult to cover back up with the leaves. Once exposed, the cereal rye continues to grow and competes with the adjacent plant. Waiting one to two weeks after applying the leaf mulch on top of the cereal rye ensures that the cereal rye is shaded

Three-fourths of an acre of garlic, onions, and potatoes mulched with leaves using a manure spreader required less than one hour of weeding over the entire season. Eight hours was required to apply the leaves with the manure spreader.

Credit: Shawn Jadrnicek

long enough to kill it. Alternatively, I planted directly into the leaf mulch on top of the cereal rye and this worked the best. The leaves were shredded which helped retain moisture and the older leaves worked better than the younger leaves. Irrigating twice was enough to keep the plants alive during the wet spring we had while conducting the trials. A dedicated drip line for the plants would have alleviated any drought stress I observed.

Cold damage was the biggest issue with crops planted into the early crimped cereal rye with leaf mulch placed on top. The mulch prevented the soil from warming and the kale and broccoli I planted into the system were damaged even though they are cold hardy crops. I didn't harden the transplants off before transplanting because this hasn't been an issue when planting into bare soil, but should definitely be done with early crimped systems. The transplants placed directly into the soil under the mulch had more cold damage than the transplants placed into the mulch on top of the soil. Cold air probably settled into the depressions in the mulch made when planting into the ground. In addition, the dark-colored older leaves applied as a mulch showed less cold damage on transplants than the newer leaves, probably because the darker color and denser mulch absorbed more solar energy.

Credit: Shawn Jadrnicek

No-till broccoli planted into cereal rye terminated early by applying leaves after crimping.

Insects, Mulch and Early Crimping

Crops planted in the early spring are notoriously susceptible to cabbage maggots, a tiny fly that lays its eggs on the soil near plants. The fly larvae decimate the root system and girdle the stems, leading to a dead plant after a few days. The problem is worse when planting into a freshly tilled cover crop. The freshly decomposing organic matter attracts the flies and the larvae move from dead cover crops to transplants. To combat the problem, I only plant early spring crops into a field of winter-killed cowpeas. The cowpeas decompose through the winter and in early spring no fresh organic matter is left to attract and feed cabbage maggots.

With the early crimped cereal rye covered by leaf mulch, I expected cabbage maggots to thrive in the fresh decomposing cereal rye below the leaves, consuming any transplants in the system. To my surprise, I had no cabbage maggot damage in the treatments. Early one morning while observing the trials, I noticed swarms of fungus gnats hovering above the leaf mulch on top of the cereal rye. As the fungus gnats landed on the mulch to lay eggs, a pack of predatory mites would emerge from the mulch and consume the eggs immediately. I'm assuming the predatory mites played a role in controlling the cabbage maggot problem, allowing a massive amount of fresh cereal rye organic matter to be applied in the early spring.

Delaying Cereal Rye Maturity

Another happy accident occurred when I conducted the trials with leaf mulch covering the cereal rye. Every treatment in my plots had a control that was crimped but not covered with leaf mulch. Cereal rye in these areas regrew from the early crimping operation. However, I noticed the cereal rye matured later in these areas. Late maturing cereal rye is useful for late-planted crops in no-till systems. Since crimped cereal rye only prevents weeds for six weeks, it's important to plant soon after crimping. If you

No-till kale planted into cereal rye terminated early by applying leaves after crimping.

Credit: Shawn Jadrnicek

crimp when the cereal rye is mature, then wait two weeks to plant, only four weeks of weed control are left.

With late-planted crops, four weeks of weed control may not be enough to get the cash crops established. Crimping cereal rye to delay maturity provides a solution, but it must be done at the correct time. If done too late, some of the cereal rye will stay down and decompose leaving thin areas in the mulching system. More research needs to be done in this area. However, from my observations it appears that crimping when the cereal rye was two feet in height at the pre-heading stage allows maximum regrowth while still delaying maturity by around two weeks.

Irrigation

Within our intensive cover cropping systems, I've seen the organic matter content of the soil increase 0.5 percent every year; it now reads close to 6 percent. I strive to use no-till techniques as much as possible and follow cash crops with cover crops by avoiding double- and triple-cropping systems. Now that we're adding mulch to the crimped cover crop systems, I expect organic matter contents to climb much faster. I've read that for every 1 percent increase in organic matter, the soil can hold up to one inch more water. More water stored in the soil means more opportunities to dry farm or reduce irrigation. In the potato and garlic systems described earlier, I only irrigate at the critical crop development stages. When planting winter squash into crimped cover crops, irrigating a few times after transplanting helps plants establish enough to dry farm during a normal rain year on the East Coast. All other crops are irrigated with drip tape placed on top of the crimped cover crop. Pre-irrigating with the drip tape makes planting by hand much easier if the soil is dry. Transplants are then easily watered in and automatically irrigated with the drip system.

Fertility

Cover crops rarely provide all the fertility to grow the vegetable crops unless legumes are added. Therefore, with rye only cover crops crimped as a mulch, I add fertilizer to the system. I do this prior to planting or I band the fertilizer beside the plants after transplanting. I've successfully used blood meal and cotton seed meal with surface-applied techniques. However, during a dry year, the fertilizer may not wash down into the soil to feed the plants and I've observed deficiencies during drought. I believe fertigation systems injecting the fertilizer through the drip tape would be ideal, but I've yet to adopt the practice. Another consideration with fertility is the increase in organic matter content over time. Calculations indicate mineralization of the organic matter in our soils is releasing 75 pounds of nitrogen a year. While not all is released during a short cropping period, the organic matter is still delivering a substantial amount of fertility to the crops.

Conclusion

The trials I conducted were not analyzed in a scientific way, just through observations taken during my free time. More research is needed to verify my observations and expand the potential for the techniques. Analyzing wood chip and compost mulch in addition to leaf mulch on top of a wide range of cover crops at different depths and at different times for early crimping and extended weed control would help advance and verify the techniques even more. With all the new organic no-till techniques, I see a future where 75 percent of our crops are grown using no-till techniques—saving us time and money while improving the soil.

I TOLD MY FRIEND JULIA SHANKS, AUTHOR OF *The Farmer's Office*, that I was working on a book about no-till growing methods. She told me about this no-till farm in Orange, Massachusetts, that I should check out called Seeds of Solidarity. Before I even had the chance to contact them, a letter came in the mail from Ricky Baruc, the farmer.

Ricky had been getting *Growing for Market* magazine, and read the stories we were running about no-till. He had been developing his own no-till system for the past twenty years. The letter had clippings of articles he had written for other publications, asking if I'd like an article about his system.

I called Ricky back and told him I was interested in an article for the magazine (which ran in February 2018), and asked him if I could interview him for this book, which is how I found myself on his farm on a hot summer day in June 2017.

I wanted to talk with Ricky because his method used silage tarps and compost as a mulch like some of the other systems I had seen, but incorporated cardboard as a mulch in a way I hadn't seen. I realized this was a way to deal with the bane of no-till systems—perennial weeds. Ricky's method is a way to establish growing beds and build soil, almost anywhere, regardless of whether you have good soil or perennial weeds. Which jives with Seeds of Solidarity's motto to "grow food everywhere."

Seeds of Solidarity Farm

Ricky Baruc and Deb Habib
Orange, Massachusetts
Mixed vegetables
Occultation, cardboard and other mulches

Origin of Seeds of Solidarity Farm

Thirty years previously, Ricky had been farming twenty acres in New York State. "It was organic, but not sustainable both for the use of fossil fuel and high level of personal burnout. I lost my joy of farming and never thought I would go back into it for my livelihood," said Ricky.

"In the '80s when I got into farming, the standard was 20–30 acres. Things have changed from when we were looking for farmland. When we were looking back in the day, we wanted river bottomland. Now that's getting hundred-year floods three years in a row. So now, people are going up in the hills.

"That was the typical scale, and so you would have that equipment. There are still people that want to go that route, but there are so many young people that don't have access to land and don't even have interest in the tractors and stuff. They just want to grow food. I wish I knew the numbers, but there have got to be thousands of people. Even people doing it on a small scale, they're still tilling. You don't have to.

"It's about thinking smarter. There are more and more people asking, how do you build this land? How does the farm get richer each year rather than getting depleted? That's why so many people are going towards reduced tillage."

When they moved to their current location, Ricky and his wife Deb Habib founded an educational nonprofit, the Seeds of Solidarity Education Center. The emphasis for both of them is the intersection of social justice and food issues, to promote their goal to "awaken the power of youth, schools, and families to Grow Food Everywhere to transform hunger to health, and create resilient lives and communities."

When they realized the education center would only support one of them full-time, Ricky decided to try farming their new piece of land, despite the fact that people told him it couldn't be farmed. The farm is on the kind of land deemed "not suitable for agriculture" because it's too steep, too rocky, and the soils aren't good enough. When they arrived on the land, there was nothing but a small clearing that had been used to yard logs.

Thinking conventionally in terms of flat expanses you can drive a tractor on wasn't going to work. Their land couldn't be farmed like that. But by using occultation and cardboard, Ricky

Spring garlic emerges in well-mulched permanent raised beds.

Credit: Seeds of Solidarity

was able to carve fields out and build soil that now grows great crops. His is a good example for someone looking to break in raw or forested land without a lot of machinery.

"How do you grow where the people are? That's the thing. You're looking at pieces of land that aren't suitable for equipment. How do you grow intensively between buildings, whatever the case is? You've got to be innovative. We're flexible," said Ricky.

No-till opens land up that isn't good for cultivation to vegetable and flower growing. "Buying good farmland is difficult, because you're competing with development for that good, flat land, and there's only so much of it left," said Ricky.

"The article I'll write some day will be 'farming without farmland.' These methods are also applicable to urban areas, which is critical. If you're trying to grow food, and you're thinking you've got to buy farmland, you've got to buy equipment, or buy a piece of land with a house on it and a barn, who's going to be able to do it? We want people to get into it, but the lens to see this through, is that there was no land here," said Ricky. "This was forest."

"When we came here, I was so burnt out from farming. I had a choice. So I got back to farming on this little land we had." The adversity of farming on woods and pasture that had been abandoned for years required developing different techniques from what Ricky was used to using.

As we walk around the farm, we visit fields that have been carved out of the woods on the flattest land the property has to offer. Ricky has about two acres of fields, with a half acre here and a quarter acre there.

"For that first field we saw up there, we went by a co-op, loaded up the car with boxes, and we just started putting down cardboard," said Ricky.

"That's the other piece of this, is how do we grow food, without [prime] land, with minimal labor, without getting so stressed out? On so many farms when they get it tilled the weeds go crazy. The silage tarps are great. Basically, when we got on the land here, there was no land here. These fields were used around the Civil War," said Ricky

We're addicted to constantly adding fertilizer, but how come these farmers were farming for thousands of years without fertilizer?

— Ricky Baruc

Credit: Andrew Mefferd

Corn was planted into compost on top of cardboard. Ricky is digging down through the compost to find the cardboard layer.

gesturing to the fields closest to the road, "but down [by the house] there was no land.

"We had to build everything. I had no land to do cover crops, now I have enough land and the silage tarps allow me to get clear land to put in cover crops. Now the mix is, you've got the silage tarps, you got the cover crops, I got rye and vetch growing in a whole bunch of places, and my favorite is the cardboard," said Ricky as I followed him to his lower fields. "Cover crops, silage tarps, permanent beds, and then the cardboard. That's what I'm after. I'll show you more about that down below.

"Basically, we're talking no land, no equipment, minimal labor, so how are you going to farm? You can't find good labor around here. This way I can basically farm alone. I had apprentices. Some years it was great, some years miserable. What I realized is, when I was working by myself, it's like, 'I'm happy right now. Huh.' You end up managing the group and you're never satisfied with their work, right?"

"So where did the inspiration and ideas come from? Tell me about the process where you came to this system," I asked.

"In Montague, a few towns over, I had quite a big market garden. I knew about black plastic. I assumed the black plastic would kill sod. I don't even know how it came to me, but I just said, 'What about cardboard?' Sometimes the best idea is one you don't mull over, it just happens. So I just went for it. And when we moved here, I needed to open up land. I just started taking out cardboard. I didn't know, there was no guarantee," said Ricky.

"The silage tarp was because of the lamb's quarters disaster. That became my second tool. And the biggest thing was that I had absolutely no mechanical aptitude. I just have no interest in spending any of my time doing that."

Ricky's enthusiasm for the system is visible as we walk around the farm. We walk past broccoli, kale, beets, onions, greens, chard, all kinds of crops planted in permanent beds.

"That's the thing I'm interested in is that—we don't know what's going on in the soil, and there's a lot going on. If I can inspire people that it's not only better for not releasing CO_2, there are so many benefits to not tilling. I think that's going to be the sales pitch," said Ricky.

Choosing Between Occultation and Cardboard

"Basically, my whole thing is if you're going to do transplants, use cardboard. If you need to plant cover crops or scatter sow seeds or direct sow, use the silage tarps. That's how I distinguish the [methods]," said Ricky.

"Then you can do all your transplants into the cardboard. You put your cover crop or scatter sow seeds where the silage tarps were. Then as soon as you can when you have bare ground, put in cover crops. You have cardboard, silage tarps, cover crops. Those are the three things.

"Stress is only going to increase with this crazy weather. You want to figure out how to minimize the stress not knowing if there's going to be a drought, or heavy rains, you just don't know. It's modern-day growing. How do we grow in this crazy erratic climate? I think that's the beauty of both methods, is they're keeping the moisture in, but also we found in years we had too much moisture, it doesn't allow too much moisture to saturate the cardboard, as well as the silage tarps. It gets a lot of stuff in, but it doesn't let the moisture leave. It keeps a steady state of moisture, which is good."

Cardboard Mulch

The most concise explanation of Ricky's cardboard mulch method appeared in the article he wrote for *Growing for Market* in February 2018. So

We find lots of worms in the compost and cardboard.

Credit: Andrew Mefferd

here it is, followed by the discussion we had while looking at it on his farm in 2017:

Cardboard is my way to open up new land, my weed control, moisture control, and worm food that results in nutrient rich castings—my primary fertilizer. One full pickup truck load of large sheets of cardboard from furniture, bike, or appliance stores will cover an area 35 by 100 feet.

Let's say it is early spring, even with a bit of snow still on the ground. I open large cardboard boxes with a box cutter so that it is one layer thick. Then I lay the sheets on my field, ensuring that there is at least a three-inch overlap between pieces. You can remove any plastic tape now, but it is also fun to do so when that is the only thing left!

The soil that he has built is so loose and friable Ricky can easily dig with his hand up to his elbow.

Credit: Andrew Mefferd

I have my covering material nearby, be it mulch hay or inexpensive manure—partially decomposed horse manure is fine for this purpose. The mulch is to hold the cardboard in place and keep it moist. It is a good idea to cover as you go, so the wind doesn't pick up the pieces. I lay cardboard right over my permanent raised beds; it will conform to their shape once it gets wet and begins to soften. If you know your soil is not very fertile, put an inch or two of finished compost or manure down before you lay the cardboard down to increase fertility as well as speed the decomposition of the cardboard with the infusion of worms and microbes.

Cardboard is a great method to use for transplants. In as little as two to three weeks, the cardboard will be soft enough to dibble through for planting seedlings. I grow a lot of bunching greens like kale and collards, and also use this technique with success for garlic, corn, and squash. At the same time, there is no rush—the cardboard will be your mulch, keeping weeds at bay and the life in the soil happy until you are ready to plant. This is also a great technique for fields that you wish to leave fallow to replenish; once the cardboard is fully decomposed months later, you can decide to cover crop or add more cardboard.

In the areas where I have been using cardboard for at least five years, the growing beds have become so established and rich that I no longer need additional compost and have never added lime; the cardboard and subsequent worm, microbial and mycorrhizal activity balances the soil and feeds the plants. I call it the 'no-till self-sustaining cardboard method.' This is huge in regards to the cost of acquiring compost or other fertilizers, as well as labor time and cost.

A common question that people ask is whether cardboard is safe. Our experience is that the use of plain brown corrugated cardboard (the hide glue used is high in protein, which also attracts the worms) promotes incredible life in the soil. A telling year of the moisture moderating benefits of cardboard came in 2016 during that incredible season of drought. Two inches below the cardboard and without irrigation, the soil was moist and worms were active, and above, the crops were fine." [Author's note: Some organic certifiers prohibit cardboard that has ink in colors other than black. As always, check with your certifier if you are certified organic.]

The soil underneath the cardboard is rich, black, and loose.

Credit: Andrew Mefferd

"The cardboard is the key for the worm population," Ricky explains. "It really hit me last year. During the drought I was able to go under the cardboard and the soil under there was moist and full of worms. You look at things politically, spiritually, and environmentally. The political way to look at it is tilling is like this culture. It's like we're going to go in there even though this incredible relationship in the soil has developed, we're going to go in there and just [destroy everything]."

"It's a destructive release. You do release fertility but at the expense of all those relationships," I said.

"There's an initial rush, that's what the farmers have liked, but it's burning out the savings account. Every time you till you're taking savings out of the account. People like Elaine Ingham and Michael Phillips, they're all writing about this stuff. That's why this no-till book is critical at this time. Once people figure out what's going on in the soil, they're not going to want to mess with living soil. How do you do that? How do we grow food in a way that's minimally disturbing?" said Ricky.

The residue from the previous crop is still visible, though it is dry and there are no weeds.

"Remember, there was no soil here, so this has been built up over the years. We just went in there, covered it with cardboard. Then we got the cheap compost, just to weigh it down, and for some nutrients. Let that cardboard get wet and I dibbled through it and so the beauty of it is all that labor saved, plus weed control, moisture control. And then that cardboard is the fertilizer, so ultimately what I'm after is the *Farmers of Forty Centuries* stuff," said Ricky. [Author's note: *Farmers of Forty Centuries* is a book that looks at traditional farming methods and how fertility was maintained over long periods of time; see the Resources section.]

Credit: Andrew Mefferd

"We're addicted to constantly adding fertilizer, but how come these farmers were farming for thousands of years without fertilizer? I think we're used to adding this amount of fertilizer based on a soil that's not very alive. Can you get a real living soil by not tilling and minimal inputs? That's what I'm most interested in, with the input of cardboard. So it's multifunction as it covers the land so I don't have to till, it's the weed control, it's the moisture control, and it's the fertilizer. That's what I'm after. I'm really going after minimal inputs on not very good land," said Ricky.

"The corn looks pretty good. We'll see what this looks like underneath. This is the stuff that I just love doing," said Ricky, kneeling down and pulling back the mulch at the base of a bed of corn.

"It's been dry, but I just love the moisture. I think in terms of growing, that plants are like humans. They want it moderated. They don't like it too hot, too wet. That's the nice thing about the mulch. It keeps it pretty steady. This is what we're after, look at that right there," Ricky says as he scoops out soil that is moist, crumbly, and filled with worms.

"What we tell people is, cardboard is the perfect worm food. Worm poop is the perfect plant food. This is from Amy Stewart's book [*The Earth Moved: On the Remarkable Achievements of Earthworms*]. One acre of land per year with heavy worm populations will produce 150 tons of worm castings. You can see why. They go year-round. The only time worms are in jeopardy is if you get an early frost. One nice thing about the cardboard is it moderates the soil so [you're protected from extremes]. You can see they were really hanging out where it was the most moist," said Ricky.

"Oh, yeah," I say, "worms are all over the place. This is all built soil?"

"Yes. So rather than incorporating in the soil by tilling it, there is this design to deal with it, just like the forest floor. [Organic matter goes on top,] microorganisms will come up and break it down," said Ricky.

"That's beautiful soil," I said as Ricky dug down through the earth with his hands. The soil was dark and crumbly like potting soil.

Meanwhile, Ricky continued making the hole bigger. "Hey, look at that. That's the base. When we brought in sand to level, that's where it started. You can see that's sand," said Ricky, having scooped a foot or more deep down to a sandy layer with his hands. In the greenhouses there is as much as two feet of soil on top of the sand.

"There was no soil. What we tell people is, you don't have to till it. [This amount of topsoil is what you get] twenty years later. Well, maybe not quite that many, but you get the idea.

"The cardboard I think is really a winner. As you can see, [the earthworms are] doing the work. Cardboard is my favorite for transplanting into. Hands down, it's really building soil. Silage tarps are my second favorite because they allow me not to have to incorporate the [residue], whatever it is, weeds or cover crops. Or to have to hoe and take it away, so it keeps the organic matter right there. I wonder to myself, all of our cover crops once you till them in, what detriment is that, the tilling part?" said Ricky.

We look at an area with a dense stand of winter rye on it.

"This was a corn crop that was in cardboard. And the nice thing about the cardboard, it really hangs in there long enough [for long-term weed suppression]. There really wasn't much left. We just had to pull out the corn stubble, very minimal weeding, and I was able to rake and then just put the rye in."

"So this rye was planted into a layer of compost that was on top of cardboard from the previous corn crop?" I asked.

In hoophouse tomatoes planted through cardboard, the root zone stays moist even though it is hot and dry.

Credit: Andrew Mefferd

"Well, the cardboard breaks down pretty quickly. There was cardboard, compost, corn stalks. And so this spring, I got rid of the corn stubble and just planted right into it. Yes, so no manipulation," said Ricky.

In response to my question, Ricky started digging around in the soil under the rye with his hands. After getting down through the soil, he found the remains of the cardboard.

"Look at this. There's a little cardboard left, which is interesting. The corn stubble is still here as you can see," said Ricky.

We walked into a hoophouse that had a spinach crop in it.

"This is a good little example here. A spinach crop, did great. Now the crop is done. What is a person to do? You could get your tiller in here and till the whole thing. I come in here and cover it with cardboard. You could use silage tarps, but in this case I'm going to do cardboard. A silage tarp is perfect when you're going to direct sow. Cardboard is perfect for transplanting into. That's really the distinction. Because in the greenhouses I plant mainly greens that I scatter sow, like Bryan O'Hara (see interview p. 305). I want bare ground. I'll put compost on top and then scatter sow seeds," said Ricky.

"I knew in here I was going to transplant. I came in here real quick. Imagine the amount of time it would take me to till in that spinach, wait for [it to break down], and then who knows what is happening to the soil because we don't really understand soil at this point. Instead, I put down the cardboard that day or a couple days later, whatever, cut out the holes and put compost in the holes and planted tomatoes into it. Then all that organic matter is dying and I haven't disturbed anything, with minimal weeding.

"Thinking about the farmer, the crop's done, and you've got all this other stuff happening. Now I got to deal with this stuff. This way I can just throw the cardboard down. It's saving me lots of labor. [If you tilled,] you're still going to have to either make beds again or weed. This way it's multipurpose."

Ricky started digging around under the cardboard the tomatoes were transplanted into. He showed me the layer below the cardboard where the spinach was. The previous crop of spinach is all gone.

"Here you go. It was all spring spinach. Now there is no sign of spinach. That's just magical," said Ricky.

"And how much time passed between when you put the cardboard on and when you transplanted your tomatoes?" I asked.

"I'd be doing it that day. Yeah, that's the beauty of it. With tilling, farmers love that tilling because it releases all the fertility, right? You're getting that with this thing, too, because all [the residue from the previous crop is] breaking down and feeding [the next crop]," said Ricky.

Silage Tarps/Occultation

Here's Ricky's description of how he uses silage tarps, also from the February 2018 *Growing for Market*:

> When I want bare ground to sow cover crops or scatter-sow seed for salad greens, such as in our hoophouses, I use silage covers. I use large silage covers to create darkness to rapidly turn fresh biomass into mulch and leave bare soil below.
>
> Seedlings thrive in cardboard mulch, but to sow seeds directly without tilling, silage covers (also called bunker or

panda covers) are a great method. These are much thicker and more durable than regular black plastic, last many years, and can be easily moved around a farm or garden as needed. Weeds or cover crops under the tarp die and become mulch that is transplanted into, or raked off and composted in the paths.

Two months after placing a silage cover down, all the weeds have become a layer of broken-down biomass that I can transplant right into without needing to till or remake raised beds, an enormous labor saver! Weed seeds germinate in the warm, moist conditions generated by the tarp, but are then killed by the absence of light.

Once tillage stops, the cycle of bringing weed seeds up from below is broken. The beauty of silage covers is that the previous crop becomes a carpet of biomass that will create a barrier to many of the un-germinated weed seeds still in the bed. And as long as I add clean compost above the biomass into which I then plant, I have minimal weed issues.

I also use the silage covers in our four, 30-×-96-foot hoophouses, where we do "cut and come again" greens production in succession through the season. I place the covers over the beds of greens after they have provided multiple harvests but are no longer marketable. The heat of the hoophouse really speeds up the breakdown of the crop residue below. All the green matter becomes brown and I cover this biomass with a half-inch layer of compost and sow the next crop of salad greens right into that.

Or, I can easily rake off the broken-down biomass whereas before I had to hoe the pre-

Garlic grows weed free in heavily mulched beds.

Credit: Andrew Mefferd

vious crop and remove it. I was somewhat concerned as to whether the silage covers would harm the life in the soil. So one day I looked under all of the dead biomass of the previous greens crop, after two weeks under a silage cover. The biomass was moist and the soil below it was cool, creating a perfect environment for worms to do what they love: eat, reproduce and poop, with a wonderful new layer of castings produced.

Silage tarps give me breathing room. Once the season is going, there are so many pieces of farming that demand attention. To be able to cover a field once crops have been harvested, wait until the plant material breaks down, then decide how to best use an area eases my mind, versus looking around, seeing weeds and stressing.

I have also found that silage tarps are a great way to incorporate cover crops into the soil, rather than tilling them in. This technique would be helpful to those who do use machinery during a wet spring, when you can't get on the land to till in an overwintered cover crop. Additionally, land that is covered with a silage tarp in early spring will help the land to warm up more quickly for planting, as well as activating the soil to nourish your plants to come.

Besides tremendous soil building and other environmentally beneficial aspects, no-till helps with the stress factor of farming. There can be great stress associated with the feeling of "I've got to get these fields tilled now," and then once tilled 'I've got to get those plants or seeds in the ground before the weeds get a head start.' When I see all of our farmland planted or mulched in cardboard, regenerating with silage tarps, or in cover crops, I feel all is well and cared for.

For those out there who would like to use their hands more, and machinery less or not at all, decrease labor due to fewer weeds, reduce fertilizer inputs and eliminate release of CO_2 from tilling, these techniques may be for you. It is heartening to see a movement towards mimicking nature, as evidenced by the many farming articles and conference presentations that focus on revering the life in the soil. As farmers, it is critical to keep ourselves healthy for the long haul, and leave land better than it was when we started working it. After farming for thirty years, the practices I've developed, arrived at, continue to explore, and love to share nourish me as a human being by creating an honoring relationship with the soil, one that enhances life.

Having seen mulches applied to suppress weeds in no-till systems on other farms, I asked Ricky how he came up with the method. Like so many inventions, Ricky's came from dealing with adversity.

"We had this great chicken compost we were using for years. And then a bunch of years back, I was planting three 100-foot beds a week. And two weeks into it, it was all just lamb's quarters. So somehow, either here or at the farm, lamb's quarters inoculated all the compost," said Ricky. "The point is, how do I get rid of them? I found that FarmTek had this heavy-duty silage tarp. So I used that to kill all of the weeds."

"Then I went to this farmer [meeting for] reduced tillage, I brought this concept of silage tarps. And they were laughing at me, I kid you not. A guy at the meeting said he Googled it, and 'There's no word *occultation*.' They just thought I was totally nuts.

"But Cornell is doing research with these silage tarps, and what's happening with them, is that it's heating up the soil in the spring, which is really critical for getting the nutrients available. The other really cool thing about the silage tarps is, cover crops are great as you know, but if you have a wet spring, you can't get on the land, and then you have to till in this cover crop, and you have to wait a period of time [for the cover crop to break down in the soil. By then] the season's basically over."

"Yeah, in the northeast US it is," I said.

"These days you're almost guaranteed a wet spring. At least with permanent beds, you don't have to be stressed that you can't get the tractor on. And if you do have cover crops, using these silage tarps at least you can get the cover crops out of the way in the springtime without manipulating the soil. There's no reason you have to manipulate the soil with the tarps," said Ricky.

We look at a field that developed a weed problem after the previous crop was done. "I didn't have enough silage tarps and within a month it was up like this with weeds and stuff. I kid you not, two-foot-plus weeds. And I did a workshop here, it was a great demo to say, look. The beauty of this thing [is that the tarps killed the overgrown weeds without tillage]. Now I just pull the tarp back and I can plant bed by bed, so I'm not stressing out. Say you have a garlic crop, and then the garlic crop comes out, there's going to be weeds. You cover it, you let it sit, and that's where the whole Jean-Martin Fortier thing is. If you just time things right [you can let the tarps sit and kill the weeds while you do other things]," said Ricky.

Ricky was not the first no-tiller I had talked with to bring up Jean-Martin Fortier. In fact, many of them that used tarps to kill weeds referenced Jean-Martin, and as far as I can tell he's the reason growers in North America are calling tarping "occultation"; I certainly hadn't heard the term until I read his book. To see what everyone is talking about, read the sidebars quoted from Jean-Martin's book *The Market Gardener*.

Tarps and Pre-Crop Ground Cover

One of the most important discoveries we made throughout the years has been that of relying on soil-covering tarps to smother crop debris when preparing new ground.

Until then, our only way of clearing the remains of finished crops and established weeds was to either till them into the ground with multiple passes of the rototiller or manually

From Jean-Martin Fortier, *The Market Gardener*, p. 50.

Credit: Andrew Mefferd

Permanent beds under occultation in a garlic field.

remove them. As we started to move away from relying on the tiller, we often favored hand picking out weeds and residues, rationalizing this time-consuming activity by telling ourselves that all this organic material we were bringing to the compost pile would eventually become great soil building material. This way of working was labor-intensive and time-consuming, and our beds were never really cleaned of the smaller weeds.

Then one midsummer's day, I bought a big black UV-treated polyethylene tarp to break open new ground where we planned to plant berries. My idea was to dry it out before putting it away in the shed. As it happened, it stayed there for three weeks, and when we finally moved it away, it struck us—the tarps had killed all crop residues and weeds, leaving us with a very clean bed surface to work on. We had stumbled onto a technique that was highly effective. Ever since, we have used tarps to cover the ground as a complement to our minimal tillage system.

Every time a harvest is done with, we immediately cover it. Depending on what crop is next in line on the schedule, the tarp will remain on the bed for anywhere between two to four weeks, leaving our minds worry-free. Passively, we are preparing the soil for the next seeding while also weeding it just like a false seedbed would.

Over time we have consistently noticed a difference in weed pressure on beds that have been covered compared with others that have not. The explanation is simple: the tarp creates warm moist conditions in which weed seeds germinate, but the young weeds are then killed by the absence of light. Looking into this we found out that French growers were widely using this technique (called occultation) to diminish, or even eliminate, weed infestations in their fields.

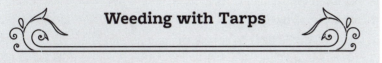

Weeding with Tarps

The main factor in keeping a garden weed-free is how much space is to be kept under control. Our 10 plots have a total area of 1½ acres, and if we had to cultivate the whole garden every week, I doubt we would manage. This is where the opaque UV-treated tarps come in handy. Not only are they useful for smothering weedy ground in preparing the soil before planting, but when covering unused beds, the tarp limits the surface area on which weeds can establish themselves. Even more interestingly we have also observed that black tarps do an especially good job of diminishing weed pressure on subsequent crops....

We have been using 6 mm black silage tarps in the garden for almost a decade now and I can say without hesitation that their usefulness is one of the reasons behind the overall success of our operation. This passive and efficient practice takes care of part of the weeding chores while we are getting work done elsewhere in the garden. Besides being a petroleum product, the only difficulty we face using these bulky tarps is that they are heavy to move around. Our solution to this has been to buy more of them every year, with the intended goal of having one for every plot, thus eliminating the need to carry them from one garden to another. This minor challenge aside, the overall advantages far outweigh the drawbacks.

From Jean-Martin Fortier, *The Market Gardener*, p. 104–05

"With the silage tarp, what kind of time frame are we talking about for leaving it down?" I asked.

"In the greenhouse, we're talking three weeks; outside, a month, month and a half. The longer the better, for sure. In this one greenhouse you'll see we had early greens. When they were done, I just let them keep growing because I knew I didn't have to freak out. Because in the old days I'd have to scythe them, hoe them, haul [the biomass] out. Talk about labor, right? Now basically all I have to do is scythe, cover, done. I can let them grow as tall as I want. I knew I wasn't going to replant there for a while," said Ricky.

Occultation acts as a placeholder so unplanted beds don't grow weeds in this hoophouse.

Credit: Andrew Mefferd

"In the greenhouse, I was concerned, was I killing the worm population? What's happening is all that organic matter is on the top, the worms are happy in the moisture underneath and so they're not being bothered."

We walk over to another one of Ricky's greenhouses. A silage tarp is covering the whole growing area, black side up. Ricky pulls up a corner to reveal the remains of a tall crop, which is flat and light green. It reminds me of hay before it is baled.

"Is this the leftover crop or is this a cover crop?" I asked.

"Good question. This was my first round of spring greens, but they've all gone to seed. I wanted to bring bees here, so I let them flower. Three-foot-high plants all going to seed big time," said Ricky.

"This was all salad mix, spinach, arugula, lettuce heads. I started early this year, so this was my first crop. Now I'll be coming back in here and getting the second crop. This is beautiful. I have enough space so I didn't need this immediately. This is the rotation, in a sense. Look, the moisture is still there. It didn't dry out, right?"

"That's a good point. If it were uncovered it would be really dry right now," I said. Even though the remains of the crop aren't succulent anymore, the soil under the tarp is still moist.

"I was smart enough to remember to write down when we put this silage tarp down. It went down on May 30, what's today?" Ricky asked.

"June 13," I replied.

"All right, this is 14 days later. May 30, 2017, silage tarp down. Two weeks later, my choice is this. I could put compost on and either plant greens or transplant into it. I don't usually in the greenhouses, but I could put cardboard down. Or look at this, look how easy it is to do this," said Ricky.

While we're talking, Ricky is raking the dried plant residue off of the bed. Even though it flowered and grew a lot more biomass than most greens crops ever do, after two weeks under the silage tarp inside the hoophouse, the remnants of the crop are easy to remove from the top of the bed.

"Look how easy that is," he says. "This whole thing, [the bolted greens] were maybe three feet high. Imagine what I used to do before. Because back in the day, I was cranking the greens out to lots of restaurants, so I had to really time it all carefully. I'd come in here and scythe and hoe and I'd bring it over there [to the compost pile]. That was a lot of labor.

"Look how easy this comes out. You can either rake this off if you need it clear. Most likely at this point I would just put compost right over it. What I'm trying to figure out is how to do things with the least amount of labor. Because why move it away if it's going to make organic matter here?"

Under the silage tarp, under the partially decomposed greens crop that had gone to flower, the soil is still moist and friable. Seeing this with Ricky, I realized that he basically turned his cash crop (salad

Credit: Seeds of Solidarity

Time is saved with the silage tarp turning weeds into biomass; a field is ready to replant without having to till or remake raised beds.

greens) into a cover crop (by letting it go to flower, something most growers would avoid at all costs), both giving the bees a source of nectar early in the season and giving himself a cover crop without having to till or plant anything else beyond the main crop.

"So I put this tarp down. Here we are two weeks later. Now I can just basically unroll and start planting," said Ricky.

"I put a temperature probe outside in the spring under the tarp. It was five or ten degrees warmer [than the uncovered soil temperature]. Actually, under the cardboard it was warmer than the normal soil too. Because sometimes you think cardboard is going to keep the soil cooler, but it wasn't cooler than the normal soil. Cardboard was warmer than the normal soil and the silage tarp was quite a bit warmer. It's ben-

No-till garlic.

Credit: Andrew Mefferd

eficial in the spring for the nutrient availability, but they're finding out it's even better in the fall too."

"I love this stuff that you're showing me. I love the fact that you can just go in and put the tarp down to terminate the previous crop without tilling or taking all that organic matter out of there," I said.

"Exactly, but even once you've tilled, then you're still stressed about what to do now. The ground is tilled and now you're even more stressed," said Ricky.

"You got to bed back up again. Re-fertilize. I don't know which is worse, but you know what I'm getting at. Either way you're stressed to get it tilled and then you're stressed once it's tilled. We take something we love, turn it into a business, and then we hate it because it stresses us out," said Ricky.

"Did you scythe the bolted spinach or did you just put the cardboard down right on top of it?" I asked.

"You could just knock it down if you didn't want to scythe it, but that stuff was really tall. There was so much of it so I scythed it," said Ricky.

Ricky does workshops on his farm demonstrating his methods. "It's so interesting when people come to the workshops, I love to hear what they say first because that's what really hit them, right? I ran into a few people from a workshop we had done some time later. What got people so interested is that we harvested the crop, and we replanted it the same day. Not to say that's the end all, but the point is that you could do that, versus getting all this equipment out and stuff. The whole tilling thing, I think, is passé," said Ricky.

We peek under another silage tarp.

"Here's a perfect example. Once again the silage tarps; I just pull it back, it's relatively clean. What's actually there is some residual garlic, so I have to hoe that out, but it's not too bad. It definitely kills anything. This is a good demo. I've never really done it before. This was planted in October," said Ricky.

Credit: Andrew Mefferd

Salad greens are planted intensively in hoophouses; previous crops are composted in place with silage tarps speeding the process.

"Look at the worms, look at that. There was cardboard there. Now the question is, where the heck is the cardboard? We're talking November, December, January, February, March, April, May, June, say seven months and the cardboard is gone. Where did it go? What sort of fertility did it give? It's not much of a weed barrier anymore as you can see, but the other thing I like about the silage tarps and this stuff is that if the weeds never get to see the sun, they won't grow. If I have a silage tarp down, I'll pull it off, immediately get compost on, plant or mulch or cardboard for that matter, it never sees the sun so the weeds are much better."

Advice for Getting Started with the System

"For people who are going into this, is that your recommendation? Throw the cardboard down, take soil out of the planting hole, and put in compost? Or if you're going to break in a new space, would you amend the area first? Is some compost right in the planting hole going to carry the plant through the season, and then you're going to build the soil up over time?" I asked.

"So these are the options. One, if you're going to plant, say, tomatoes, eggplant, peppers, bigger squashes, do what I mentioned. Put down cardboard, take the soil out of the planting hole, [fill the planting hole with] compost, and hopefully that lawn has some soil. That's your first season," said Ricky.

"Your other option is to put down a layer of good compost, cardboard on top of it, make it wet and do transplants like kale, collards, lettuce, stuff you can plant close together. Then I would do that method.

"The third method is, especially if you have contaminated soil, put maybe wood chips down and make a barrier. You can put cardboard down, good compost, and then scatter sow greens seeds, because the greens don't need much soil. Those are the three methods. Tomatoes, peppers, eggplant, big plants, you can plant maybe two, two-and-a-half feet apart, through the hole. Things you can plant 18 inches apart are the kales, lettuce, [bunching greens]. It always helps to have either grass clippings, hay, wood chips or something on top of the cardboard."

"To keep it from blowing away?" I asked.

"Yeah, or you could use [low-grade] compost just to hold it down. The thing about the cardboard, which is so great, is it's guaranteed weed control and you don't have to keep watering. The cardboard has got to be covered because otherwise it will dry out. You always want to keep something on top of it. Once it's covered, it's very mucousy, it's got hide glue in it, it's high in protein. That's what the worms are attracted to. Once they start messing with it, you're good," said Ricky.

"For the fourth technique I would go with the silage tarp. You can get any size silage tarp up to 80 by 100, which is huge.

Credit: Andrew Mefferd

"There are always going to be naysayers, but I think what's cool about a book, because it does not exist, and [the no-till methods], they're all coming from different angles. There's so many different ways of doing it. It's not like it's got to be this one way. People will get to relate to different farms and their personal stories. It's going to be so cool to have a book so people have access to the people doing no-till. Then it will keep growing.

"I think all those pieces, labor, CO_2 emissions, go into it. But it's the stress factor. We need young people to go into it. Because if nothing else it's about feeding ourselves healthy food, connecting back to the Earth. That's basically in all the spiritual books what it's all about. We've got to reconnect. We're interrelated, coming from the heart. If

A former logged area is transformed into a productive small farm landscape using cardboard and other no-till methods.

we're going to survive the next so many years, it's not just techniques; it's not just making a living. It's about reconnecting."

After a delicious lunch of tortillas and farm-grown salad, I helped Ricky and his crew pull a large silage tarp down the road onto the next plot. I said my goodbyes, turned off the recording and turned to get in my car when Ricky left me with one last thought: "If you can do it here you can do it anywhere!"

ANDREW SCHWERIN AND HIS WIFE MADELEINE grow mixed vegetables and berries on Sycamore Bend Farm in Eureka Springs in northwest Arkansas, close to Fayetteville. They have about an acre of permanent vegetable beds that are four feet wide and 100 to 150 feet long, plus a greenhouse and a high tunnel. I caught up with Andrew over the phone in the spring of 2018 because I wanted to hear how he managed his deep straw mulch system.

"I had moved to Eureka Springs, and I had this long-term plan to start a plant nursery. Around the second year that I was here, I was working odd jobs and building up plants, and a friend introduced me to Patrice [Gros's Foundation] Farm. I just felt that I needed to learn farming, and it happened to be no-till," said Andrew.

"It sounds like you're doing a system that's similar to what Patrice was doing at his farm on your own farm. Are you using his system, or did you make modifications to it?" I asked.

"My system is very close to Patrice's. I'd say the most significant difference would be that I don't have grass paths, which is a pretty minor detail. At the same time, Patrice brought a lot of ideas from Dripping Springs Garden [also in Arkansas]. And of course, then they got a lot of stuff from the basic Eliot Coleman methods of how to do intensive gardening. So it's all evolved, and taken different pathways," said Andrew.

"Would you describe the methods that you're using now?" I asked.

"To summarize, it would be straw mulch and weed management. Usually, people have to till because there are too many weeds, right? If I have, say, a lettuce crop and I harvest all the lettuce, and then I have an empty bed, we're going to have a few weeds in there. In that case, if there's spring lettuce, and I was going to plant sweet potatoes, which I will be doing this year, I would use black plastic. Occultation is another name for it," said Andrew.

Andrew Schwerin
Eureka Springs, Arkansas
Mixed vegetables, berries, and lamb
Deep straw mulch, occultation

"Have you gotten to the point where your weed pressure is pretty low, and you can go from one crop to another without weeding or cultivating?" I asked.

"I think so. Overall, weeds are a very low-priority problem for us. It's still more than what the workers want to handle. We do hand weeding for things that we don't have mulch on. For example, the carrots would be one," said Andrew.

"We would just hand weed, or with a little Japanese hand hoe, to get the weeds in between the carrots. On a larger scale, a person could still cultivate that, without a tiller, just using a finger weeder or something like that. There are a lot of options out there."

"So have you been able to work your weed seed bank down so there just aren't that many weeds germinating anymore?" I asked.

"Yes, that's what I'm going for. And I believe I'm doing that. I don't know if all my workers agree with that belief. But yes, I'm very adamant about not letting any weeds go to seed in the garden," said Andrew.

Sweet potatoes to the left, fall garden on the right.

Credit: Andrew Schwerin

"There are weeds like henbit, which aren't so bad. One we do have continual problems with is carpetweed. That one takes a lot of our attention, and dandelions. I've given up on dandelions this year. I've kept them out of the garden for four years, by spending a lot of time taking them out, around the garden, but they win."

The Deep-Mulch Method

"I know that Patrice used a lot of straw mulch, almost as a form of occultation to keep the soil covered and keep it from growing weeds over the winter. Is that what you do?" I asked.

"Yes, absolutely," said Andrew.

"And so, you leave everything covered, and you have permanent beds, and you just pull

Credit: Andrew Schwerin

What the beginning of the season looks like for sweet potatoes in Sycamore Bend's system.

Credit: Andrew Schwerin

This is what rolling straw mulch out in the fall on Sycamore Bend Farm looks like.

Theresa follows diggers; washing then clipping and flipping.

Six foot bed spacing is completely hidden in August, controlling most weeds.

Credit: Andrew Schwerin

You can see how the straw is suppressing the weeds before the sweet potatoes overgrow and start smothering weeds.

the straw back to expose fairly weed-free soil when it's time to plant?" I asked.

"Exactly. So this time of year [spring], we're usually trying to pull the mulch back about three days before we're seeding, or transplanting, just to let the soil warm up a little bit more; because it stays cooler underneath the mulch. But we don't want to do it earlier, because we don't want the weed seeds to get a head start. And in case we have heavy rains, to avoid erosion," said Andrew.

"So you must be buying a lot of straw?" I asked.

"I reckon it's probably about $750 a year, for one acre. I use large, round bales instead of the square bales that Patrice was using. A round bale covers about 200 feet," said Andrew.

"So you're pulling the straw back to expose the bed for planting. Are you leaving the straw in the pathways, or do you mulch back around the crop during the growing season?" I asked.

"If it's a long-term crop, we bring it into the pathway, and it's there for, usually, a few weeks. Then we bring it back into the bed. For crops that we are planting closely, if I'm doing something like radishes, or if I'm transplanting lettuce, the straw doesn't go back into the bed, because those crops are so densely planted," said Andrew.

"If it's something like tomatoes or peppers, we don't take the straw all the way off the bed. We just dig a little hole, where the plant is going to go, and we plant the seedling through the hole in the straw. And then, there's a big middle ground. Kale and broccoli are ones where we mostly clear the bed. We transplant, as they get bigger, three weeks down the road, then we re-mulch them. So they're mostly weed free for the season."

"What are the challenges with the deep straw mulch system?" I asked.

"The main thing with the straw is removing it. And this is mostly for scaling it up; I don't consider it a problem here. If it's not going to go back in the beds, we don't want to just leave it in the pathways, because it can get piled too high. This is where I have this idea, I wish I could find out if somebody has tried it, of running a hay baler through [to pick up] the straw, and then you have bales of hay at the end, which you can move elsewhere," said Andrew.

"So often we will make a big pile at the end of the bed, and a few months down the road, we'll need that, and we'll move it somewhere else in the garden. But that's a lot of labor. It's not a lot at our scale, but

Summer kale showing weed suppression.

Credit: Andrew Schwerin

it would be a large amount if you were doing seven acres, or something like that.

"It doesn't take any longer to roll out a bale of straw than it does to get the tiller out and go up and down the beds with it. People have this need to go into their garden in the springtime and get rid of all the weeds, and I don't think they're taking into account the cost of cultivation, when they say that the straw costs too much. Because it would save a lot of money by using it, and it's easier to save money than make money on a farm."

Recommendations for Getting Started

"What's your advice for somebody trying to get started with no till?" I asked.

"My recommendation would be to use straw mulch, and start it the year before, or at least, the fall before they want to plant. If you're putting it on in the spring, it's just going to be really hard to take care of it. You're going to have too many weeds, which are already established, coming through. So it really has to get on before winter starts," said Andrew.

Credit: Andrew Schwerin

Mulched weed control in greens crops.

"That makes sense. I've heard similar advice, to get started ahead of time with no-till, from other people," I said.

"If they're using plastic [for occultation or solarization], there's still the issue of what happens when you take the plastic off to plant. Then those weeds are going to come fast. So the mulch is good for being able to continually keep the weeds down. I'm really pro-straw, instead of anti-till," said Andrew.

"I know you said that you run the rototiller over your beds, just really shallowly, almost using it like a Tilther to fluff up the soil surface, from time to time. What does that do? Why do you still do that little amount of rototilling?" I asked.

"It's to incorporate raw manure. A lot of the nitrogen will [volatilize] if it's not buried right away. So I like to get, at least, dust covering

Credit: Andrew Schwerin

Summer lettuce, with
lots of tomatoes and
cucumbers, squash,
potatoes, and other greens.

it, so that it can [get in] the soil immediately, instead of sitting on top, drying out," said Andrew.

"Can you estimate how deeply you run that rototiller?" I asked.

"Since it's not tilled much, it doesn't go that deep. About one inch. I don't have to do that all the time. For example, if I'm putting the manure on in the fall, then I don't need to till, if I'm putting the straw on right after the manure," said Andrew.

"Is there anything else I should have asked you, or anything else you want to say about no-till?" I asked.

"Your big question was, difficulties and successes, and I'd say, over-all, it's successful. I'm happy with this. Maybe you should ask me in midsummer. I've never thought of changing, and it's not because I'm stubborn. It's just working," said Andrew.

I spoke with Dan Heryer of Urbavore Farm in Kansas City, Missouri, late in November 2017. I had heard about Dan and wife Brooke Selvaggio's urban farming for a number of years. What I didn't realize was that they were doing it without tilling, until someone said, "Well, have you talked to Dan and Brooke?"

After neighbors opposed their original in-city farm, they moved to a 13-acre parcel in the city and established a diversified farm with fruit trees, livestock, and no-till vegetables. That was a few years ago, so I wanted to check in and see how starting their farm over with no-till was going. The farm they built on scraped-off soil (including a former baseball diamond) inside city limits is unique and fascinating, and too complicated to fully detail here.

Dan Heryer and Brooke Selvaggio
Kansas City, Missouri
Mixed vegetables
Applied organic mulch/straw

A number of articles have been written about Urbavore Farm's transformations over the years, and they are well worth reading. We are going to focus pretty tightly in on their no-till system. To hear about the rest of their farm, listen to the Farmer to Farmer podcast they were on, or seek out one of the many articles about them.

Starting out

"Why did you start with no-till and how did you come up with a method?" I asked.

"Shortly after Brooke and I first met a little over ten years ago, I started working as the farm manager for a local farm/nonprofit, which is now called Cultivate Kansas City. Back then it was Kansas City Urban Agriculture," said Dan.

"When I was working for them, I attended a local growers conference for the Midwest region, and there just happened to be this guy from Arkansas, Patrice Gros, who has Foundation Farm. He spoke about his no-till method and how it was for him as far as the tilth-building, reduced labor, and the various aspects. His was a deep mulch system," said Dan.

"I was very inspired by him and that year I managed the farm at Cultivate Kansas City. There was a guy there who had done your typical market gardening, and he was into trying other methods. He and I were interested in this no-till method, so we trialed a few beds in no-till and liked the results.

"The next year I started farming with Brooke, so we decided to just go for it and try to do no-till. At the time she was farming her grandfather's backyard. It was probably an acre and a half in suburbia here in Kansas City.

"We decided to try doing no-till on the whole thing. On her grandfather's land, we didn't have any expenses, so we were able to take big risks. We did it over the course of two seasons and immediately ran into problems in terms of efficiencies and things like that. There were plants that were way too wet and cold, and that threw off all of our planting timing, everything that we had known.

"We started trying to figure out that learning curve. We were committed to it and we just kept going with it. Ours was a heavy organic matter, deep mulch sort of method at that point. We were using mulch on the beds that were previously tilled, and then we were using heavy, thick mulch to kill weeds. That was where we started out at, and did a lot of transplanting, and moved on from there."

Origins of Urbavore Farm and the No-Till System

"The piece that you point out about figuring out how to do things with really low costs and as minimal investment as possible is something that really resonates with us. We did not want to get into debt over our farm. We felt like getting into debt was the surest way to fail. So we just were adamant that we were going to try to stay out of debt," said Dan.

"We picked out the sites that had what seemed to be decent soil and reasonable slopes to work with, and those are what became the vegetable fields. We started out with four fields that were each a third of an acre. When we laid out our first field we had to look for mulch. We have a big brewery here in Kansas City called Boulevard Brewery. They were filtering their beer through these biodegradable filters and they were approximately three-and-a-half-foot squares. It was probably seven feet by three-and-a-half feet and they folded it up into squares, so there was a seam in the middle. So we would unfold these and make a row out of them, make a bed out of them essentially, and lay these down as sheet mulch," said Dan.

"Then we would put straw on top of them. By themselves they really wouldn't have been enough to kill established fescue and things like that, so we put straw on top of them, and it was a way of reducing the amount of straw we were using.

"We laid out the first plot like that and we left. We tried to leave grass strips between the rows, which was a bad idea. It was taking us too long to lay the rows out on that first field, so we decided to nix the aisles and just mulch the whole thing on the next field.

"Then at some point we left out the beer filters as well. It was a pain to try and get them here in one piece, so we decided to forget the beer filters and just lay the straw very thick, and that worked fine.

Dan in the tomatoes and flowers.

"By that time we had decided to keep the area in between rows of crops mulched. So we would clear the mulch off a four-foot area, expose the soil, and seed the whole bed of beets, for instance, or greens or something like that, anything that would be direct seeded. For the first season that's a huge undertaking because you have such thick straw in there. We have four-foot beds and two-foot aisles. Some things like garlic, we'll just plant a big block of garlic and won't do an aisle.

"We went and cleared off that four-foot section for the bed and you'd end up with such an absurd amount of mulch in the two-foot aisle that it was just impossible to deal with. We're trying to adapt it to a more commercial scale, where we can really plant efficiently and not

Here's a row of onions transplanted into a break in the mulch.

Enjoying the no-till chard.

mess around with the straw too much. We want to do whatever is the most efficient without tilling, so we clear off beds and we will then have to weed those beds and manage them really well to make sure that weeds are at bay.

"There are an awful lot of things that we don't direct seed, or that don't require a full bed clearing. We just keep those in rows or seed them, open up a hole for more widely spaced stuff like summer squash or melons or stuff like that.

"Before we do anything, we'll have a plan as to what's going to be planted in the field, like most farmers would. We know what the layout of the field is ahead of time, and then we pound in T-posts and mark the edges of each bed, marking the aisles as well. Then any beds that are getting cleared off, we're going to run a twine line from post to post and clear off that bed of straw and get down to bare soil with those twine lines as guides for what we're doing.

"It's the same thing for any other crop. We mark things out and T-post the beds. They'll get pulled later and all those T-posts will then get thrown into tomato trellising and things like that. But that's how we start out.

"With the melons and cucurbits it's the same thing. Cucumbers, we would clear off two rows in a single bed instead of cleaning off the whole bed, because you want to keep things mulched so that the beds don't get overtaken with weeds. We're only going to do two rows of cucumbers in a four-foot area. We don't like to space them any tighter than that, so we clear off two lines and they'll probably be about nine inches wide. Just a straight line of bare soil. With the cucumbers we furrow a little bit in the soil for them straight down the row, seed them, and cover them with compost, like we do most of our direct-seeded crops.

"Now for something like summer squash, we do a single row per bed, two feet apart. So we have a measuring tape down, and run down the line and open up the holes first, and then take the measuring tape off and put down compost. Bring the tractor bucket over with the compost, scoop five-gallon pails out of the tractor bucket, and then go down the row. A bucket will probably do five or six holes, so that's a lot of compost per hole.

"Then we would seed into that compost. Now depending on what kind of weather conditions we're expecting, we might just put the seed in the compost with no soil contact, or we may flatten out the compost pile down there and have direct soil contact with the compost on top of it. So we're poking the seed down accordingly, depending on how much moisture we're expecting, because the compost can be quite dry and then sometimes it doesn't germinate. We don't irrigate at all, seeds are never going to see water other than rain. So if the compost is too dry, then [seeds won't germinate]."

"But you smash it down if it's going to be dry?" I asked.

"Yes. Our first planting, we probably just go ahead and do it in the compost and not do it on the soil. But in subsequent plantings of summer squash we would probably assume that it was going to be dry and press for soil contact. Just that little bit of moisture on the surface of the soil is going to be enough to germinate the seeds," said Dan.

"Then the transplanted crops work much the same way as for the summer squash. We put out the measuring tape, open up holes accordingly in the mulch, and then put our compost in the holes, and then we would take a trowel, or

No-till furrows for direct seeding.

Credit: Brooke Selvaggio

A chicken tractor after the vegetable crop has come out.

No-till kale and nasturtiums.

in the case of tomatoes we'll take a shovel, and dig a hole. Not a deep hole, but it's just more efficient with a shovel with the tomato hole.

"And we do our perennials the same sort of way. We are definitely in a transition with how we do things. We have had a huge amount of problems with perennial weeds, particularly bindweed and milkweed, so we are trying to figure out how to deal with that.

"We have some four-year-old plantings of strawberries that are getting kind of tired and they need to be replanted, so we are laying out an area for that as well. It's been a while since we laid out a new area in sod. So we're going to lay silage tarps on that. We have a couple of used billboard [covers] that we keep around for doing solarization in some areas when it gets out of control. Generally we don't use tarps, but we're moving in the direction of trying to use a little less mulch. We're still using plenty of mulch, but we're using tarps to reduce that need to some extent."

No-Till Nordell-Style Cover Crops

"Are you familiar with [Anne and Eric] Nordell?" said Dan. "I really admire their cover cropping system and the way that they've reduced weed pressure on their farm. Obviously, the way that they've accomplished that is through tillage, and we want to mimic their cover cropping system with no-till. We want to try to cover crop at different times of the year and use the pigs to turn under those cover crops, and as a result kill different types of weed seeds over the course of many seasons.

"One of the things that's important to their system is their fallow periods, where they let weed seeds germinate and then till them after they've germinated. That is something that is tricky to accomplish with no-till. So we're hoping that once the pigs run an area, that we can use silage tarps to bare fallow, essentially. To let the weeds come up after the pigs are off, then tarp it and kill those weeds, and then either plant another cover crop or let the mulch sit and prepare for the next season.

"Our hope is that the pigs will be a game-changer as far as scalability goes. We hope to use them on areas that do need weed management. It certainly happens, especially with areas that we cover crop, where things get weedy and need to be dealt with in some fashion," said Dan.

"Your basic options are to re-mulch it to a level that it's going to kill it, or to hoe it up, which is what we do. We typically will just surface hoe any areas that get really weedy, and then we're probably just hoeing the top inch of the soil. But it's really prohibitive in terms of labor. Our hope is to turn things over without mechanical tillage, but with these animals. We did that with the chickens to some extent, but it can't be too weedy for the chickens, because they can only handle so much."

Fertility and Long-Term Planning

"On the Farmer to Farmer podcast you said to start off by putting a foot or a foot and a half, a lot of straw down. When you're breaking in new ground, do you put down any fertilizer or compost or anything?" I asked.

"When we initially got here we limed the whole 13 acres, which without equipment was a huge pain. What we do for fertility is animal based, so we run the chickens. If there's an area that we want to prep and fertilize for growing, then we will run the chickens on that area heavily and take the manure from their houses and spread it throughout the plot that we're planning on mulching. And that's the main source of fertility," said Dan.

"Then the compost at planting supplements that, but we don't just broadcast compost. What we use in our field we buy from a local company that produces finished compost. It is also food-based, so it's pretty hot."

"If you're going to make a new patch you need to be thinking about it the previous year. There's no equivalent in your system of plowing it up quickly, right?" I asked.

"It does definitely take long-term planning. You have to know what you're doing. But we have the established plots, and those are a lot more flexible than a scenario where we want to plow new ground or something like that to put a crop in real quick. We can't do that. We have to plan probably a year in advance and think, 'Okay, we want to plant strawberries next spring, so we're going to run the chickens heavily in the summer and then in the fall we're going to mulch it and by spring it'll be ready to plant,'" said Dan.

Harvesting no-till muskmelons.

Credit: Brooke Selvaggio

Farming the Unfarmable, Without Irrigation

"We couldn't farm that baseball diamond if we didn't do no-till. If you plowed that up and tilled it, it would be a crusty clay mess and you couldn't grow healthy crops in it. But our single best tomato crop that we've had on this farm was in that baseball area, and you just couldn't do that in a tilled soil situation," said Dan.

"We definitely have our problems, but our crops are really nutrient dense, really flavorful. The difference in the quality of the crops on all levels, as far as taste and aesthetics, is noticeable and noted by our customers. I don't know exactly what's happening underneath that straw, but I know it's good. I know it works.

"In addition to the benefits of low cost as far as tractor and implements and startup and all that sort of stuff, the quality of the food that comes from a natural system is second to none. It's really just very, very healthy, especially over the course of time. The plants get healthier. The soil gets healthier. Things are improved.

"There are problems. The bindweed is a huge problem, but it doesn't really impact the health of the crops until it starts climbing on them. They're still very happy and healthy living with the bindweed right next them. It doesn't negate the fact that the health of the soil and the quality of the produce is absolutely worth whatever problems are there.

"We don't irrigate and I think that is a major benefit. Irrigation costs a lot of money and a lot of resources. I was writing something for the website development that we're trying to do last night and 80 percent of the water in the United States goes to agriculture. We don't use any, so that's huge.

Dan transplanting into mulched beds with his son on his back.

"Our climate is not like California. It's not super dry like that. We have about 38 inches of rainfall annually on average, but we went through a very severe drought in 2012, where from the beginning of the year we probably had 13 or 14 inches until the end of August. That was an extremely dry year. Things were crunching under your feet.

"It was dry, and we still produced about $50,000 worth of crops, in our second year here. We lost stuff for sure, but we saw other farmers pouring irrigation into their fields to keep things alive. They were pouring thousands of dollars worth of water onto their fields and still failing at it, while we had crops on our table. So it's significant what is happening underneath the mulch and what is going on with the soil building and the structure of the soil."

"It sounds like overall though you're going to stick with no-till?" I asked.

"For sure. One of the other problems that occasionally happens with no-till is too much water. Especially when you first lay down a new plot, things can get very wet and it can be problematic. I've seen people try no-till deep mulch systems on their farms and they think, 'Whoa, things are way too wet. I have too much mulch on everything.' I think that, first of all, is very much a problem initially when you start the system, and can be a problem in a really wet season. Certainly there are areas of our farm that we have taken out of production because they're too wet. I don't know for sure whether or not they would be too wet if we were tilling them," said Dan.

"In a no-till system they're too wet for us and we need to address those sort of issues in those areas, if we ever are going to crop them. It'll definitely reveal where water is on your land and it can be a drawback in that respect. We've been limited to the three fields that we're currently growing on because of the water issues that have come up in other areas of the farm. So it's worth noting."

On the day I visited Bare Mountain Farm in Shedd, Oregon, it seemed like the state was trying to live up to its reputation for drizzly weather. It was mid-October, and Denise and Tony Gaetz invited me in for a cup of coffee while we waited to see if the rain would let up enough that we could go outside and look at the flowers.

BARE MOUNTAIN FARM

Denise and Tony Gaetz
Shedd, Oregon
Cut flowers
Occultation and compost application

When Megan and Jonathan Leiss of Spring Forth Farm in North Carolina (see profile p. 295) told me some of their systems were modeled after Bare Mountain's, I wanted to make sure to get out and interview the Gaetzes. If you look them up on YouTube, Bare Mountain Farm has a bunch of videos where they talk about what they do, and give really great practical information on how to do it.

I visited Denise and Tony on the day of their potential first freeze, as the weather threatened to turn colder that night. They were pretty relaxed, though, ready to let some things go if they got zapped, knowing that other crops would make it through in their high tunnels.

The Gaetzes grow flowers on about two-and-a-half acres. Their field production uses permanent beds with grass pathways, covered with low tunnels when necessary. They also have several high tunnels, giving them about four thousand square feet of permanently protected cropping space.

The Gaetzes are in Linn County, self-described as "the grass seed capital of the world" due to the high concentration of grass seed farms there. Though they used to do farmers markets, they now sell exclusively to florist/designers in Portland and Eugene, Oregon.

As we sat down for coffee, Denise was on the phone, trying to market the last of the season's bounty. "Our season's winding down, and I'm trying to get the last bit of product out that I can," she said.

"Normally we use an electronic store we have set up on our website. But since our availability now is so low, we're just connecting with customers on a one-on-one basis," said Tony.

Out-Competing Weeds

"Our theory is that we'll plant things very intensely and the plants will out-compete the weeds. Now, there will still be weeds in there. We don't care about it too much. We do have some spots where we're trying to eradicate some perennial weeds that got a head start on us. We have a peaty section that we're trying to rehabilitate that has Canadian thistle in it," said Tony.

"What we have found is, if you tarp those things down, you can start getting them under control. But, unfortunately, when you have a perennial weed like Canadian thistle it's a lot more manual work; we've got to rogue it out. Keep coming at it and keep coming at it.

"But for our annual beds, we've removed Canadian thistle, bindweed, and stuff like that simply by tarping. For example, we'll take a bed that we want to bring online that maybe has got a perennial weed

Late fall zinnias.

problem or something like that. We'll tarp the area for four weeks or so. That'll knock it down; it may not kill it. Then, particularly if it's the warm season, we'll plant buckwheat, which grows fast. And then, even though the thistles start to sprout up, we'll just tarp the buckwheat down again. We'll do that about three [times]. By that point, most of the root reserves are pretty much wiped out.

"And there are people I know that have been trying to get a handle on perennial weeds that are out in a larger area, and use Sudan grass. It's the same idea, except they won't tarp it. It grows real fast. Then they'll mow it down, leave the refuse on the field, irrigate it in, and it's just like successive cycles; when you come in and mow it down you're also mowing down the thistle. So you keep mowing it down, and Sudan grass will last the whole season, but then it usually winter kills. At least here in the north.

"And so, that's the whole idea of exhausting the weeds. Our theory on perennial weeds is you've just got to exhaust the root reserves, and some things are more resistant to it than others. Bindweed's probably the worst."

Armor on the Soil

"That's a refrain I've heard from no-till growers, that perennial weeds are hard to deal with, since you can't keep them from coming up by leaving them undisturbed, like you can by leaving weed seeds undisturbed in the soil," I said.

"We recognize that all around us are grasses and weeds; this is grass country. As an example, we never had Dutch white clover on the place till a couple of grass seed farms decided to put it into their rotation. Next thing we know, because there are starlings and all kinds of birds, they'll go eat the seeds, come flying back over our place and: Boom. We get the same thing with blackberries and other things of that nature. So to try to say that we're going to take our weed seed bank down to the lowest possible level is a waste of time," said Tony.

"We use a lot of tarps. Like, when we're finished with a crop we'll knock the crop down and drag a tarp over it. Depending on the time of the year, the biology will work real fast. It could be as fast as three weeks depending on when it is. In some cases it's four to five weeks. And then we'll rotate in either a cover crop or a cash crop. We use a lot of purchased-in compost too.

"We got a lot of these ideas from different people, like Singing Frogs Farm (see interview p. 275) and Patrice Gros of Foundation Farm in Arkansas. He's got a different take on it; uses a lot of wheat straw. But it's the same idea; that keeping the soil covered also inhibits the weeds. And it gets to a point where your organic matter is such that any weeds that do come along are real fast to dispatch with."

"You mean they just pull out easily?" I asked.

"Just pull them fast, or you use your collinear hoe. It's just like sweeping the beds, and it's fairly fast," said Tony.

"When you're planting buckwheat, are you using a seeder?" I asked.

"I use an Earthway, or I'll just broadcast it. Most of the time, I'll broadcast it, and then I'll use a rake and just scratch over it a couple of times. If I'm concerned about bird pressure, I'll put a 30% shade cloth over it just to get it germinated."

"To keep them from eating the seeds?" I asked.

"Yes, and it also retains moisture. So if you're in a dry time of the year you can wet the bed really well with irrigation and put that shade cloth over it, and that will help retain moisture; at the same time, it prevents the birds from getting in there and pecking the seeds out," said Tony.

"And that rotation is pretty fast. So if you're taking out a bed of thistle, you can wipe it out; if you started in May, you could be ready to put fall or overwintering crops in. And by then that thistle's pretty much gone.

"What we found is that just by using a silage tarp, that will retain enough moisture in there that the straw will rot down pretty fast. You can pretty much bank on about 80 percent of it being gone. And if you get to the point where the refuse hasn't totally decomposed under that tarp, all the root systems are gone. It's very fast just to come in and rake it off to the side."

Specific Flower Crops

"Flowers in general give you a little more flexibility, because you're not growing as many low-growing, fast-turnover crops. Where I can see if you're doing something in fast turnover that weed suppression is going to be hugely important," said Tony.

"I can see how if you grow some of the taller flower crops, once you get them ahead of the weeds, that they would smother and shade them," I said.

"Yes, the canopy is such that it really slows down the weeds. They don't get a chance to go to seed by the time I tarp the beds. So it keeps the weed bank down, even though it'll never be zero," said Tony.

"We grow sunflowers, which are interesting because they get a pretty thick stem on them. So, it's the same thing as corn stover, or if you raise sorghum for an ornamental, or something like that [where there is a lot of debris left at the end of the crop]. The roots will go pretty fast, and even if they're not totally absolutely gone, I don't really care. I'm going to plant into it. Usually, I'll try to target a crop that's going to start out being a little taller, so if I do have refuse left on the bed even though the roots are gone, I'm looking at that as a mulch most of the time.

"The only reason I'd rake it off is if I've got to plant something that's really a small plant. As an example would be dianthus, which when it goes into the field is typically a really little plant.

"So it depends on the context of what you're doing, which technique you would use. I understand from a [direct-seeding] standpoint, if you're using a Jang seeder or something like that, you want a clean seedbed. But a lot of our stuff is grown from transplants, and is a fairly good size by the time it's ready to go in.

"The other difference with us, why we don't sweat the weeds so much, is the concept that

The Bare Mountain Farm house viewed from the field.

Credit: Andrew Mefferd

the turns on flowers are a lot longer. There aren't any flowers that we know of that are going to have [a turnaround time] like radishes: gone in three weeks.

"We don't flip the beds that fast. We get two good rotations through a bed in a season in the field. And then three in a tunnel. And not much in the winter. Because light levels, once you get past the middle of November, drop to a point where things will grow but they won't bloom. So we look at our season as about ten months.

"We can prep ourselves ahead of time with some greens. That'll get us a head start in a tunnel, but the flowering aspect of things for us doesn't really start until mid-January up here, even in a tunnel. And then it stops around mid-November. Mums are pretty much the last thing. Sometimes they'll go to Thanksgiving, but that's about as far as we'll get."

"What are your frost-free months?" I asked.

"We're expecting there might be one today. It's hovering right around freezing. It might take out my zinnias in the field. We're going to lower the tunnels this afternoon," said Denise. "But our normal, first real frost is Halloween."

"And the last one will usually be around the end of April," said Tony.

"Typically though, the frost we get this time of year [in the fall], the first ones, they'll come in and they'll nip things or burn them back. And that ruins the quality, but it doesn't necessarily kill the plant. But the plant doesn't have enough time and daylight to actually do anything more. So once the blooms or buds are damaged it's over.

"I think a lot of people, too, when they hear 'no-till' they think of the Rodale system where you're growing rye, tamping it down and planting through that. And we've done that. It does work for larger items, like sunflowers. But there's a whole variety of other systems: using tarps, using compost, using mulch, using quick-turn cover crops; those types of things. So it depends on what you're going to be planting and what techniques you're going to be using. And I don't think that people understand the full breadth of the different tools that are out there.

"And even people who are no-till, can still use a Tilther or something like that, which technically isn't really tilling; you're just stirring up the surface maybe an inch deep or so. It doesn't really disturb the biology that much. Anything that disturbs the biology to a deep level is something we want to stay away from."

Getting Started with No-Till

"We got started in the flower business back in 2003, with a tractor and tilling. And it worked okay for a couple of years. But then the quality of the plants began to decline. The insect pressure began to just go through the roof. We had a lot more plant diseases. We were getting things planted later and later, because we couldn't get in to till because the water wasn't draining correctly. The big tractor tiller and even the small rototiller created a super-hard pan under our beds," said Tony.

Credit: Andrew Mefferd

"When we noticed this change, it didn't take twenty years, it took three to four years. It just depends on where you are, and how deep that bank of fertility is, if you may be able to get away with abusing things for longer.

"If you have a good deep soil with great structure, high organic matter, and strong mineral content, tillage damage may take longer to greatly impact yields. However, mychorrizal and other fungi are destroyed very quickly. Increased air results in higher bacterial action, faster oxidizing organic materials, and quickly reducing soil carbon. Bottom line is it's just a matter of time before one sees the need for application of more fertilizer, and higher disease and insect pressure as the biological balance of the soil is degraded."

"We just watched things decline. It went fast," said Denise. "The culmination of the frustration came when we had a field of dahlias that

Bare Mountain uses a lot of low tunnels to extend their season. The blue strip down the middle of the bed is a sprinkler hose, which delivers sprinkler irrigation to one bed at a time, useful for getting seeds to germinate where there is no overhead irrigation (or rain).

weren't thriving. They weren't really growing or blooming or anything, and we were fertilizing, trying to do everything. And I walked down and looked at them, and realized that we're not even growing weeds out here. Weeds weren't even growing!"

"Yes, that's a bad sign," I said.

"One of the weeds that was the worst when we were tilling was pigweed. We had tons of that. Now it's hardly around the place. Now it's like it doesn't fit. People have got to start paying attention to the biology of [their systems]; if you start changing that biology on the top, it'll start changing what weeds are going to be thriving. That's another thing that we've noticed has changed dramatically, and just in a very few short years," said Tony.

"The other thing is that the hard pan in our permanent beds, which are on the uphill side of our place, is gone. Totally. In some cases, we have eighteen to twenty-four inches in depth before we hit any compaction. That means all our topsoil is at least getting air and water to some degree down there.

"When we started this, we paid a lot of attention to the micronutrient analysis of the soil. And this soil was really deficient in a lot of minerals. Our calcium/magnesium ratio was way out of balance, which also creates other problems with availability of things like phosphorus.

Building and Increasing the Water-Holding Capacity of the Soil

"The other thing that is important to note, with the no till system, is it really impacts your irrigation. You don't need to irrigate as much," said Tony.

"I've heard that from a couple of other people too," I said.

"Yes. You can really cut back. We use a drip system on the place. Some overhead with some things, but I'd say 90 percent of it is on drip. And during the heat of the summer, I think the plants are even more resilient to the heat because the soil is retaining moisture so much better, instead of drying out in the first inch or so. The increased canopy, that helps too. We plant everything really intensively," said Tony.

"That's the other aspect to this place, is we're not blessed with great land."

"If you can make your own soil by layering on organic matter, and building your own soil up on top of the existing poor soil, then it should work almost anywhere," I said.

"That's essentially what we're doing here. I don't know what our soil class is, other than 'sucks in its natural state!' I make fun of it, but in reality this soil has some very good building blocks. It's not really virgin soil; it'd been farmed for grass seed for many years before we even showed up. I'd say the mineral bank and a lot of other things were exhausted on it. It's very tired soil. It's been mined, I guess maybe is a better way of saying it," said Tony.

"But when you go and you look at a soil test, the amount of clay in it is actually a positive. A lot of people look at clay, and I used to look at it the same way, like, 'Why don't I have beautiful sandy loam?' But in reality what I want is clay, and I want to bring the organic matter up tremendously. We had maybe 1 or 2 percent organic matter in the soil. And you couldn't even really see it. When you get a test it just looks like dirt: lifeless, grayish-looking clay.

A shot of just the bed of rosemary.

Credit: Andrew Mefferd

"What we're attempting to do is bring that organic matter up to 6 to 8 percent, or maybe even better. And in some of our more permanent beds, where we've been doing this longer, organic matter now is up to 4 to 5 percent. And the tilth is radically changed. The other reason we like clay is because once you liberate it from that tendency to plate up and just become [rock], it has tremendous cation-exchange capacity, which is huge."

"It holds a lot of nutrients," I said.

"It holds a lot of nutrients and a lot of water. And I think that is massively important for fertility. I've heard of other people that have sandy soil, and they're actually applying bentonite clay or something like that, to get that same [nutrient and water holding] effect. That's something we don't have to worry about. The tilth of the soil has great potential of improving dramatically. It's becoming incredibly fertile soil," said Tony.

Learning from Others

"We haven't actually met any of these people who are doing no-till. Thank God for the internet and YouTube and people's willingness to share. We've learned a lot and have been able to experiment with a lot too. What it comes down to is 'keeping armor on the soil.' That's a term that Gabe Brown uses, who runs a big farm in North Dakota. I don't know if you've seen any of his videos," said Tony.

"Yes, I have," I said. (Gabe Brown is well worth checking out if you're interested in no-till row crops on a larger scale).

"Ray Archuleta has some great resources out there on soil health too. They're ranchers, but basically it's the same concepts. It's keeping armor on the soil and building up that organic matter. If you keep that biology alive and going as long as you can during the course of the year, every year it just gets better and better. I think that people are kind of missing the boat on that," said Tony.

"The other thing that we're trying to get away with, with the two of us basically running this thing, we want to let nature do more things."

"We call it our ground peeps. Let the ground peeps do the work," said Denise.

"What we're trying to do, too, is reduce complexity. The more complexity, the more it costs you because it's going to cost you in time. Time is money. And you want some free time for yourself. The more complexity you have, the more steps you have to do, the more time it takes. That's one aspect. When people look at our place, they go, 'Wow, this does not look like a super well-run thing,' but it's amazing how much productivity you can get out of something that looks a little on the wild side," said Tony.

Tunnels and Profitability

"We get really good productivity out of the tunnels. If you calculate a dollar per square foot in the tunnels, we average probably over the course of the season, about twenty bucks a square foot," said Tony.

"Wow," I said. "That's really good."

"In the field, in the places where we're planting this year, I'd say our average is about four-and-a-half dollars a square foot, which is actually really pretty good for being out in the field. The one thing about flowers, even though they take longer and you can't get as many rotations, you tend to get a higher value per square foot. It's kind of a trade-off. When you take the Curtis Stone [author of *The Urban Farmer*] model, he says, 'It's about rapid crop rotations.'" said Tony.

"The only two crops that are really fast are sunflowers and millets. Those are the only ones that you can kind of tuck in after other crops. Most everything else is ninety days or so," said Denise.

"Yes. The [sunflowers and millet are] about sixty days. The other thing that we want to do that we haven't done successfully, and I like this idea from Singing Frogs Farm, is their hedgerow idea. I think that adds a lot of diversity and habitat potential for predators that will take care of other nasties on the farm. I've got to hand it to Paul and Elizabeth Kaiser, they've created a really cool place that is biologically super-sustainable," said Tony.

"We've got to come up with systems where nature does the heavy lifting around here. We'll concentrate on the stuff like getting it harvested and getting it to a customer. The more things that take us away from spending hours on a piece of equipment, the better."

No-Till Planting Timing

"What you'll notice is, when you get your no-till bed systems working, that it's not going to matter if it's raining or not," said Tony.

"It's whether you want to be out there or not," said Denise.

"Because the ground will drain very well, it's not going to be an issue, particularly if you're using raised beds," said Tony.

"That's something people don't realize; again, you're taking complexity out of it. You go all the way back to *The One-Straw Revolution*. The whole point of [Masanobu] Fukuoka's book was, he was trying to say when you look at what you're doing, you've got to look at the context, and try to continually look at the process and take out steps that don't make sense. Do only things that need to be done. Do them as efficiently as possible. His whole philosophy was continually removing things that didn't need to be done. It's the same idea as the Toyota manufacturing system. Eliminate *muda*. The two types of muda, those that you can't get away from that you have to do, but you try to minimize those, and the muda that you can eliminate immediately. It's just wasted steps," said Tony.

The fall fields of Bare Mountain Farm, showing most of their hoophouses. As crops are picked out, they are covered for the winter.

Credit: Andrew Mefferd

"And load leveling calendars for efficient planning. That's key. In April and May this place becomes an absolute zoo of planting, transplanting, seeding, cutting, harvesting, selling," said Denise.

"So what we have to do is, we have to integrate more perennials, and that's how you take the pressure off of trying to move in so many annuals. I guess from a veggie standpoint that doesn't really work, but it does for us. There are more and more people doing no-till, so that means there's more innovation out there."

"This is a perception, but I think one of the resistances to no-till is that it's not a quick fix. In that when you're using this system your time frame is different. Now, I'm not talking about the individual crop, because your focus becomes less crop-centric as it becomes more soil-centric," said Tony.

Breaking New Ground

"That brings me to another question. Will you tell me what your method is for breaking new ground?" I asked.

"I'll show you an area we're going to be reclaiming. Right now, we just let the weeds grow on it. We didn't really care if they went to seed, because the way we look at it the seeds don't matter. The first step is, we're going to get a big tarp. This area is going to be approximately four thousand square feet. Sixteen two-and-a-half-foot-wide beds on a four-foot center. We're not going to do anything really fancy to it. The first thing we'll do is kill all of the stuff on the top by bringing a tarp over it, like a silage tarp. Then we'll let that sit there most of the winter," said Tony.

"Then in February, we'll start peeling it back as we want to get into things. We're going to use a lot of compost and mineral top-dressing. Probably broadfork it at that point, and just let the ground work it. Once we put the minerals and compost on it, we'll throw the tarp back over it. And we'll keep that tarp on it until we're ready to plant. We have a fairly high water table here in the winter, so there'll be adequate moisture in there for worms et cetera to do their thing. The soil should be ready to put plants into probably by April of next year.

"Then, when we get ready to plant into it, we'll roll the tarp back as needed, just keeping it mostly covered. The longer it's covered, the more it tends to work that stuff in. To kill the [vegetation] off you just need to smother it for several months. And every time we finish with a bed, we're going to apply a top-dressing of compost, probably about a half inch. We'll keep track of where our minerals are with a soil test, and fork only as necessary. If it's a small enough area we'll even use a digging fork or something like that."

Compost

"What we want to do is to be able to just lay [organic matter] in a bed, because there's a huge amount of minerals in there, and let the rotting process take it back to the soil. Instead of hauling it to a new pile, turning the pile, bringing the pile back and spreading it out. It's another one of those complexity things that we want to get rid of. For a new space we tarp it to kill that biomass off on the top. We don't even bother breaking up the sod, it doesn't matter at that point.

"If you don't have to have raised beds, just mark out your beds. I wouldn't go through the effort of making raised beds. If you have great drainage, don't bother with it. As you rotate beds, you tarp them out. Because we have the luxury of time to be able to tarp it. Apply more compost and any other fertility that needs to be there by top-dressing. We don't till it in. We just let that moisture drag the material down. Or even in the action of planting into it, we'll disturb the soil just enough that it'll push some of that stuff down there.

"You'd be surprised how fast worms and things like that, they grab minerals and they're just pulling it down all the time. I'll show you a bed when we're out there that we've top-dressed with compost for the winter and all the minerals that we put down, you can't see them. It's been two or three weeks, and the stuff goes right into the soil. I think that's probably the easiest way to do it."

"Let the worms do that for you," I said.

"Yes. I know a lot of people who think, 'Oh, I want to get started, so I'll till it first.' That's fine, but if you've got the luxury of time, why bother doing that?" said Tony.

"It does take some planning and logistics. That's kind of tricky for us. I'm always trying to tell him my seeds are coming, and I need them to go into the ground. He's always reminding me there's four weeks of that tarp there," said Denise.

"Well, you gotta look at a tarping as a rotation," said Tony.

Denise needed to do some more sales to get the week's flowers sold, so Tony and I walked out into the drizzle to look at the field.

"Here's our peony bed. There's a bit of weeds in there. Quite a few weeds, but what we're going to do is, I use an Austrian scythe. I'll come through this next month and I'll just whack that stuff down. We'll haul all the refuse off the top of it, because peonies can be subject to diseases. Just burn that off and then we'll throw these landscape fabric tarps over it for a couple of months. They don't start sprouting until March or so, so that would kill off any of the winter weeds and keep it pretty clean," said Tony.

We look at their compost pile.

"Then in the spring, once we pull the tarps back, we'll take this compost and put an inch or two down. This is really good stuff. We use a lot of this on the beds, to give us organic matter in combination with any cover crop we do. This gives us our big bang for the buck, and it's relatively easy for us to do because it is from thirty miles away. So, he can bring a truck in here and just dump it, it's reasonably priced. We get it for $22 a yard. So that's pretty cheap, and it's clean, which is amazing," said Tony.

"Yeah, that's great-looking compost. It's so black, I've hardly ever seen compost that black," I said.

"Yes, it's made from urban waste and chipped wood, so it's got a lot of wood in it, screenings that come off of flake board mills. Stuff like that they can't use, they sell it to these urban composters. They used to burn that stuff. So they'll pull the fines in. It's got a wood base to it. Which I think really helps with the fungal level," said Tony.

Credit: Andrew Mefferd

Bare Mountan Farm uses sandbags to hold landscape fabric down. Since they are often tarping beds with a lot of residue on them, they buy landscape fabric that is wider than the bed to allow for the "crown" that the residue creates in the cover.

The Tour

"Here's an example of intense planting," said Tony. "There are little weeds on the outside of the bed, but really no weed pressure on the inside. Same thing with the marigolds up there, you don't see any weeds growing. You don't need to use plastic with holes burned in it, it's just a waste of time from our perspective.

"Then down here, I'm going to end up pulling the tarp off of this and planting rye and vetch in here. This one has been tarped now for a couple of months. I'm just going to plant directly into it. See all the roots are gone, there's just nothing left here.

"What I'll do also is rye and hairy vetch into this. Then I'll tarp it again probably around first part of February, and then I can plant in these beds in March. The tunnels on these beds," said Tony, gesturing to some tall plastic hoop low tunnels spanning a single bed, "we keep these hoops going here. This allows us to put plastic over things, so we can protect it from hail. Or we've done hoops inside of hoops to overwinter. We'll put row cover on the inside with the plastic on the outside. This is basically constructed caterpillar-style, so we tie down each individual hoop.

"See this stuff, it's still got another couple more weeks to break down. It's not quite gone yet," said Tony, showing me how there was still some biomass and sprouted weeds on top of the bed. It was clear that the digestion process was still going here; the bed was not ready for planting yet.

"See the weeds? Even the weeds sprout in here and they'll die too. So this one's only about two weeks into [occultation] at this point. [The method is] not really super complicated, you just throw the tarp over the top. Hold it down with sandbags and let nature do its thing. I'll give you an example. We just tarped that about a week ago," said Tony, showing me a bed with a black tarp over it. "You see how much rain we got here in the last two days?"

"Yes, it was raining pretty hard when I drove up yesterday," I said.

"Yes, this is a classic example. This bed was sunflowers, when we finish there's refuse left on it. The moisture hasn't gone all the way through, and this bed had some weeds. But the soil itself, you can dig down pretty well, it's got good moisture in it, and the roots are disappearing. We just push the sunflowers down and let nature take its course," said Tony.

"Yes, that looks pretty fluffy down there," I said.

"This bed has a 20-inch depth before we get to any kind of resistance in the soil. This bed here is all finished up," said Tony, gesturing a bed of marigolds that had been harvested. "See, what I'll do is just come through with my scythe and knock it down, throw a tarp over it, it's done. I don't worry about it too much.

"The plants are healthy for this time of the year. I don't know how much more we'll get out of them. If you look around, yeah okay, you got weed pressure around the edge of the beds, but there's nothing really in the bed itself. You don't see weeds growing in the crop," said Tony.

"Right, the crop is pretty clean," I said.

"That's what I think people are really afraid of, if they do no-till they're going to have weeds everywhere. This is exactly the opposite. Now this is a perennial bed, and you can see I've got some rodents or something digging around in here. The whole idea here is the same thing, it's got a top-dressing of compost, about an inch or two. When you go down and you get to the soil itself, there's pretty good drainage. This will just stay here and just slowly rot over the wintertime, and we'll have very few weeds. This has been like this for two weeks, so there are no winter weeds germinating in here. It will stay weed free," said Tony.

"Because of the top-dressing of compost?" I asked.

"Yes, and the compost itself being pretty coarse material, is going to be able to take a pounding in the rain, it's not going to matter. You're not going to have a soil compaction problem or an erosion problem or anything like that," said Tony.

"What is this blue strip?" I asked.

"That was from before the rains came, I just haven't picked it up. That's actually just a sprinkler; it's a roll-out hose that has holes in it. I just did it to wet things down, get the top [of the bed wet], because I didn't want to stand here with a hose," said Tony.

"They do a good job of wetting things down in a small area. Every bed has its own irrigation system. You notice we don't use drip tape out here, what we use are micro tubes. I've been using the same stuff since 2005. These stay permanently on the bed. I'll even occultate over them."

This is potentially another efficiency of no-till: if you don't have to till up the bed, you can install more or less permanent irrigation, whether it be drip like Singing Frogs or Bare Mountain or sprinklers like Neversink.

A view inside the tunnel showing celosia in the middle. Some of the edge beds have already been harvested.

Credit: Andrew Mefferd

"By the time we get finished with this upper area, we'll have somewhere around fourteen thousand square feet. So, if you figure $4 or $5 a square foot, that's not bad for a small area."

"Yes, that's great," I said.

"When we get finished, each tunnel has about 750 square feet of bed space in it, so we're going to have five of these guys and one crate house. So that's about four thousand square feet, something like that under plastic. And everything in these tunnels is no-till. We use drip irrigation when the crop is growing, but we'll try to use that sprinkler to wet things down. The biggest struggle inside these tunnels is keeping enough moisture in the top to keep the biology going," said Tony.

No-Till Philosophy

"From a no-till standpoint, how we're going to run the fields is, it's going to be a lot of tarps, rotation of cover crops, and a lot of compost. I don't worry about these weeds. They're going to do what they're going to do. But when I work on a planted area, it's just about keeping the compost level higher, the armor on the soil high," said Tony.

"If I'm going to let something go for a while, and go fallow, I'm not going to waste my own cover crop seed. We got enough seed in the ground around here that's going to grow. I know some people say, 'Well, you're just inviting weed problems.' I don't get weed problems. Once you go back into doing it, one rotation, occultation, pretty much takes the weeds out of the first couple inches of the soil. They're still there and if left to their own devices and uncovered, they're going to make their way back up.

"But mostly, what you're going to end up with is an intense planting that's going to cover the ground pretty fast, and there'd be very little weeding. Maybe initially, you might have to go through with the collinear hoe and just kind of scratch them out. But one pass usually is good enough. Then once the plants get to size, there are very little weeds," said Tony.

"And over time what happens is the more you use a bed and the more you go through those rotations the weed pressure just goes down.

The key to it, we found, is intense planting and getting that cash crop established pretty fast. Then that tends to just eliminate the weed problems.

"The more you till, I think, the worse it gets. Like I said, we used to have pigweed pretty bad around here. Now we can hardly find it. The biggest problems we have now are perennial thistle and bindweed. But even that, if we work at it we can get it under control pretty well," said Tony.

"The other thing that happens if you've got perennial weeds, like bindweed and thistle, you're going to chop up those roots and spread them around. At least doing it this way you can keep an infested area confined and a little more manageable," said Tony.

"I don't know how this scales up. Now if somebody's running two, three, four acres, I suppose it's just a matter of labor. Maybe some better techniques and bed-layout, and stuff like that. The concepts, I think, are the same."

"We cut our tarps six feet [wide] because we found that it could have a little bit of a crown from [debris from the previous crop], as well as fit over a four and half foot bed, and still be fine."

Advice for Getting Started

"Do you have any tips or advice for someone who's thinking about either starting up or transitioning to no-till?" I asked.

"Yeah, test your soil, number one. You want to get that biology going in the soil. If your soil's out of balance, and is missing key things, you may not be able to get that biology really rocking and rolling. Because that's the key to getting rid of all the crop debris when you tarp it over, is having biologically active soil. If it's sterile, after you've knocked that stuff over, it's just going to sit there and dry out. You want to get a good fungal and bacterial balance in the soil. So having a good mineral content is key," said Tony.

"The second thing is to get more carbon in the soil as much as you can: organic material, compost. Or just quick-growing succulent cover crops like buckwheat. It depends on your area. Recognize that in a transition from conventionally tilling, maybe the soil is worn out, that you may have to apply a lot more organic material than you would think.

"You may need to break up your hardpan. Don't try to broadfork when it's dry, because you'll just be sitting there bouncing on that sucker. Let some moisture soak in. If I have a dry area that I'm working on and I want to broadfork it, I'll wet it down for two or three days at a time, with that sprinkler hose, just really soak it, and let it sit. Usually by the third day, then I can come back, and the fork just slides right in. That broadforking will help bust hardpan up and gets some air in there. Then top-dress with your minerals and your compost.

"In a newer area it would probably benefit you to do some kind of subsoil work, whether using a tractor with a subsoiler on it, or Yeomans plow, or something like that. The idea is that you're trying to bust things up and get some air in there to help with the biology."

It's All About the Biology

"The key to this whole thing, making it work, is the biology. You can use any kind of techniques you want. I mean, I use tarps, I use compost. Other people don't even bother with that, they just use cover crops and roll and crimp them. Whatever it is, it's the same idea; the concept is you're keeping armor on the soil. You're keeping it away from pounding rain and burning sun as much as you can," said Tony.

"The relationship that the soil biology has with the plants growing in it is super important. Plant roots exude sugars and carbohydrates that bacteria and fungi use as food sources. Part of the soil armor concept is to keep a diversity of plants growing in the soil through the year.

"So my advice would be: Know your soil, and your objective is to build biology. With no-till, there's strategy, and tactics. The tarps and all that stuff, that's the tactical. The strategy, the long-term goal, is to build your soil biology. And that is going to drive the whole thing. You [want] a good fungal bacterial balance; depending on what you're growing, if you're growing more perennials, you'll want it more fungal. If you're growing more annuals, you maybe want to tilt the scale more to the bacterial side. But the whole point is you're feeding that biology. Think of what you're doing here as feeding the soil, feeding the biology.

"If you can, whenever possible, leave the roots in the ground. Don't yank the plants out entirely. If you've got to take the top off, I understand that. As an example, with sunflowers, if I were to come back there and want to plant another crop in there, I would cut those things off as close to the ground as possible, and then I would apply some compost and I would just plant. Don't worry about the roots, and the root balls, and all that stuff, because they disappear in a month or two.

"So in feeding the biology, you really get that biology revved up, it's going to eat all this organic matter. If your biology is dead or lacking, it's going to really look like this doesn't work. I think people are missing that. It's not just about, 'Oh, I didn't run a tiller.' It's letting the biology do the tilling. It's all the interaction of increasing the arthropods and the worms and the beetles and all that kind of stuff. It all works together.

"The photosynthetic activity in plants produces sugars that are pushed out of the roots. They feed the soil. That is what's feeding the bacteria and the fungi. So yanking those things out is not good."

"What I always try to do is to go back to nature. I'll spend some time looking at an area and think, 'A lot of things are growing in there. Why are they growing the way they're growing?' If people spent a little more time looking at nature, they would begin to realize that nature has got redundant systems that are more complicated than people believe," said Tony.

"Absolutely," I said.

"There's a reason why everything grows where it grows. What you're trying to do is create an environment that's optimal for the crops you want to grow. But, I think you could also do no-till wrong too. I think you could ignore the soil, continue to put chemical fertilizers in, and you're going to get really bad results. A lot of people use the stale seedbed technique now, where they pre-sprout the weeds. And then they come through and torch [the sprouted weeds with a flame weeder]. That works, but then you've got to ask yourself, 'What is the system here?'" said Tony.

This comment made me think about how occultation is another form of stale seedbedding, where you sprout and kill the weeds with darkness instead of flame.

"The other thing I think people should think about when they are doing no-till," said Tony, " is the complexity of your systems. Don't try to solve a problem by adding another layer of complexity or mechanics. Okay, I'll spray this or that. Start thinking about why. What we always try to go back to is, Why? Why is this happening here? Pay attention to the biology, to the nature that's going on around you, to your crops, and watch your minerals in your soil. The whole key to it is to stop focusing on your crops so much and focus on the biology. Just get that biology humming, stuff will grow. It'll take care of itself."

"That's a really good point. Because I can see how some people, if their soil has been beaten up chemically and physically, it may have very little biological activity. They might do this for one season and be disappointed. It's going to take time, if the soil is dead, for that biology to come back," I said.

Polly and Jay Armour of Four Winds Farm in Gardiner, New York, have one of the oldest no-till systems I could find. They put away the rototiller more than twenty years ago and haven't looked back. By not tilling, the Armours have gradually eliminated the reasons most people till. They have worked their weed seed bank down so low and gotten their soil so loose and high in organic matter, it would be a step backwards to till it.

"Four Winds Farm is twenty-four acres. Four acres are used to grow vegetables and the remainder of the land and some additional rented land is used to pasture a small herd of beef cattle. The manure from the cattle is collected during the winter months, combined with horse manure trucked in from a nearby farm, and composted using an innovative forced air system that doesn't require any turning," said Jay.

The day that I visited, Jay was away on a once-in-a-lifetime voyage on a tall ship his daughter was helping to crew. Since we weren't able to meet in person, Jay and I talked about their farm over email. The passage he wrote me about the beginnings of their farm is particularly insightful.

DEEP COMPOST MULCH

FOUR WINDS FARM

Polly and Jay Armour and Jenna Kincaid
Gardiner, New York
Mixed vegetables and flowers
Biodegradable mulch/compost

First off, neither Polly nor I came from farming backgrounds. We did not inherit a farm that usually comes fully equipped with machinery. We were young. I had little savings. Polly still had student loan debt. In order to be able to buy the farm, we both needed to work off farm jobs. What money we made mostly went into paying off the mortgage. We used a loan to purchase our first tractor. I bought the rototiller at an auction for a really good price.

I got a couple of people to help me load it into my truck and one of them said to me, "You know that tiller is frozen?" Needless to say, I was probably the only person at that auction who didn't know. When I got home, I put that tiller on the tractor, oiled it up really well, put a pipe wrench on it and it wouldn't move at all. I kept at it day after day and eventually got it to move a little, then a little more, then more, then got the whole thing freed up.

Now I felt like I was a real farmer. We planted winter rye in the fall, spread sheep manure (we were raising sheep at the time) during the winter along with some cow manure from our neighbor, tilling it all in in the spring, and then planting, hoping to stay ahead of the weeds. Two things happened—the weeds grew more plentiful and the soil got harder later in the growing season, so hard that to pull larger pigweed was a back-breaking job. But we were doing what the "experts" told us we should be doing.

The soils are marginal on our farm, not good for vegetable farming, and they didn't respond well to the traditional approach. After five years of struggling with this approach, thinking that we were gradually improving our soil, we met Lee Reich [author of *Weedless Gardening*] and learned about the surplus compost at Mohonk Mountain House and thought, why not give it a try. Keep in mind we were looking for a way to control weeds. In the first year we not only saw a reduction in weed pressure, but we also found the weeds we had pulled out a lot easier. We didn't understand the "why" at the time, but realized we were on to something and continued with it.

Twenty-two years later, looking back on what we have accomplished, we have created soil on our farm incredibly high in organic matter. We see firsthand how that high organic matter soil performs, just like the textbooks say it will. It holds moisture during drought periods, meaning we have to water less. And it absorbs water during intense rain events.

Scientists tell us that one of the effects we will see from climate change is an increase in drought and intense rain events. There is information on how much water soil will hold with each percent increase of organic matter. We are only cultivating four acres. But if every farm in our watershed had soil organic matter numbers like ours, they would be a major contributor to reduced flooding in the low-lying areas of our watershed.

Polly and I both studied some form of environmental studies in college. We approached farming from an environmental position, which enabled us to be open to trying something different. If we had come from farming backgrounds, we may never have been open to trying something different.

This probably explains why tillage-based farmers are not willing to change. The people who are interested are the next wave of young farmers who are coming from non-farm backgrounds, who don't have capital to invest into machinery, who see what we do and realize a decent farm income can be earned on two or three acres of garden space.

Credit: Andrew Mefferd

The entrance to Four Winds Farm.

Four Winds Farm does three to four farmers markets a year, some wholesale, sells to restaurants, and has a big seedling sale in the spring. Polly and farm manager Jenna Kincaid showed me around the farm on a hot day in early August 2017.

The farm also supports a sixty-member CSA, which is managed separately. This is a great opportunity for journeyman growers to plug in to the Armours' system and gain experience before starting a farm of their own, and for the Armours to have a CSA on the farm without having to run it.

"We started the CSA in 1993 because that was the only way we could market our vegetables. Gradually, area farmers markets became good marketing outlets and our increased success with our annual seedling sale gave us enough income that we didn't need the income that the CSA generated anymore. In 2008, two employees came up with the idea of taking over the operation of the CSA," said Jay.

"The CSA is a separate business. They use the space, they use our equipment, and they use our techniques and so forth. It's a turnkey business. They have an assigned area each year. We put our potatoes together, just because it's a similar block of management, but other than that, they have their land and we have our land," said Polly. "This is a great way to both hand off some of the management and act as an incubator farm."

The no-till system used on Four Winds Farm is so simple, the biggest challenge for many growers would be taking the leap of faith away from tillage. As Jay told me in an email, "Several people have told me I need to write a book about my methods, but I'm no writer and my approach is so simple I think the book would be five pages long!" That's being modest but it's also true—one of the beautiful things about the Armours' system is its simplicity. It boils down to using a thick layer of compost as a mulch to block weeds from emerging.

As Polly pointed out, the toughest thing about this system might be getting started. The first year, you have as much weed pressure as you did when you were tilling, since your weed seed bank isn't worked down yet, but you don't have tillage as an option to get rid of the weeds anymore. "The key to making it work is to have a lot of compost on hand," said Jay. The Armours are living proof that sticking with the system can have huge benefits, like low weed pressure, high organic matter, harvesting and replanting the same day, growing a lot of food on small acreage, and the efficiencies of not investing time and money in tillage.

"Another really important point is that a high soil organic matter is attainable, which leads to better moisture retention, less water (and nutrient) runoff, and carbon is staying in the soil instead of being released into the atmosphere," said Jay.

A view of the beds on Four Winds Farm in summer.

Origin of the System

I asked Polly what led them to this method in the first place, over twenty years ago.

"Well, it sort of evolved over time. When I was in college in the '80s, I went to Cook College, which was the New Jersey agriculture school at the time. I was an ecology major, because they didn't have organic farming as a subject. My major professor was really keen on getting people to think in a systems sense," said Polly.

"He explained how soil works with respect to weeds, and how most of the weeds that germinate are in the top couple inches of soil. So when you till, you're constantly replenishing that seed bank, bringing deeper seeds up to the surface. If you could stop tilling, you could break that cycle, but the trick was, how?"

"A lot of people at that point were doing double-dug raised beds in their backyard garden. The question for us, once we started this farm, was, how could we scale that up to a farm-scale model? It's one thing to do it in a 10-foot by 20-foot plot in your backyard. It's something very different to do it on an acre. Then we started working with a working group that was starting up a CSA, Phillies Bridge, which is just to the

Credit: Four Winds Farm

Fall beets that were planted in the same beds that the onions in the previous photo were in.

north of here. We encountered some folks there that were doing interesting stuff. We met Lee Reich, who has written some interesting books like *Weedless Gardening*, where he talks about a similar method. We realized that, okay; maybe this could be scaled up. What we needed was a source of compost, because that's the linchpin for this.

"Mohonk Mountain House at that point had just begun composting all their food waste, and they were producing this gourmet compost and giving it away by the truckload. We got a number of truckloads, and that enabled us to transition to permanent beds."

"At the time we were tilling our garden space with a tractor-mounted rototiller, so we tilled everything in the spring, formed the beds with a back blade mounted on an old Ford 8N [tractor], and spread the compost from a garden cart. We made a few piles of compost ahead of time, brought over in the back of a pickup truck. Two years later I sold the rototiller," said Jay.

"Then, using the compost from Mohonk, we were able to put compost on top. We did this section first," Polly said, gesturing to the plot in front of us, "and the results were really amazing. It cut down on the weeding incredibly. The plants were healthier. It cut down on the labor involved. The soil developed a long-term structure, and we were hooked. Then the issue became, how can we scale this up?

"If we didn't get the compost down, weeds would come up. We were moving compost at that point with a garden cart, so the initial investment was pretty significant."

"How large was that first experiment?" I asked.

"Just half an acre," Polly said.

"And you didn't put any weed barrier or anything down?" I asked.

"Nothing like that," said Polly, "just a big layer of compost, after we had rototilled and so forth. The first year was difficult, because your weed pressure is the same [as before] that first year."

"Do you think a biodegradable weed barrier, like planter's paper or cardboard, applied underneath the compost would help to get started with this system, by suppressing some of the weeds that would tend to come through the compost? I'm just trying to think of any way to ease the first year's weed pressure," I said.

"The layer of compost is all you need. The weed seed germinates from exposure to light. Once you remove that light with the layer of compost, the weed seeds won't germinate. If you weren't tilling the soil first, the planter's paper might be a good idea, but it isn't needed with our approach," said Jay.

"Another important detail in our system is that we are using unscreened compost. Shavings from the horse manure sit on the soil at least a year helping to block the light reaching the soil beneath. I have seen other farmers attempt to duplicate our system with screened compost and it breaks down in around two months, allowing the weeds to become a problem in early August."

"That enabled us to realize that the system worked. Then we moved to the next field over there, and we converted that to raised beds. You see that field is a little bit lower than this one?" Polly said, gesturing to a field with a small pond on the edge.

"The water table is quite high in that area. And when we don't have to rototill or wait for the soil to dry out to work it, you can get on the soil any time of the year. I have a picture from that era of Jay and I working in the field. You can see standing water where that pathway is. We're

Top: Jay Armour hooked this blower up to the manifold of perforated pipes to aerate his compost pile without having to turn it.
Bottom: This old dog house protects the blower from the elements.

Credit: Four Winds Farm

Credit: Andrew Mefferd

A close-up view of the onions that are being harvested from nearly weed-free beds, with the comfrey border in the foreground.

working, and putting down compost, and preparing the soil for the growing season in March. That really woke us up to how much labor and time we save by not tilling."

Efficiencies of the System

"That's such a huge advantage. I can remember so many times on my own farm, having to wait for things to dry out, to be able to do that initial cultivation," I said.

"That alone is enough reason to do it, in my mind, because you don't have to constantly postpone. When you've got a rainy spring, and you are weeks behind," said Polly. "And it works also in the latter end of the season, because we actually prepare a lot of our fields for the following year in the fall. So we can get on them even earlier," said Polly.

"That's brilliant," I said.

"That's a wonderful time saver. You say it's a big investment to put the compost down and prepare the beds, and it is, but it's so much less than all the other stuff you would have to do if you didn't do that. It's a time saver in an absolute sense, and that time saving persists over the seasons," said Polly.

"I tell people I would much rather put the energy into putting compost or mulch down than to pulling weeds up, because when you put any kind of organic matter down as a mulch, you're adding to the soil. You're making it better, in addition to not having to weed. It's an amendment, in a sense."

Turning to another field, Polly said, "We did this field in '93, so that's more than twenty years ago, and then we did that field where you see the kale, that was the second. That was the second year, 1994. That revolutionized our ability to farm. We were only on this side of the farm at that point, only cultivating these two fields, and we thought there was no possible way we would ever get any bigger, because the two of us, we just couldn't handle it. Why would we ever want to? We would have to hire people, and all that sort of thing. Then once we saw the major benefits and labor saving for this raised-bed permanent system, we started expanding to the other side, over there, which used to be a hay field.

"And now that garden over there is twice as big as what we have here. As we expanded the growing area, we hired our neighbor to come and plow with his big machine, and he rototilled, and then we built the beds using a blade that we put on our little tractor.

"This is a nice size for us now, because first off, hiring people is great because they're spending their time picking things rather than just weeding. So their labor is so much more valuable, because it goes directly to a crop that earns money. It's also allowed us to expand not just in the amount of ground that we're growing, but also in the things that we do here. We have farmers markets that we sell at. We have a CSA. They now run independently of us, and it's a great system, because we don't have to run it anymore."

"We have four acres of established beds, but we don't need that much land to make a decent living. It makes sense to have the CSA here using some of the space," said Jay.

"It's also great for the farmer because it gives them a sort of journeyman farm opportunity, where they can try running a farm. It gives people a chance to hone their management skills, so we've done that for more than ten years now. Usually our interns from the previous year, or from a previous year, become the CSA farmers at some point. That's worked really well," said Polly.

No-till tomatoes mulched with oat straw to help reduce soil splash and the diseases that come with it. They are interplanted every other row with a low growing crop to increase airflow and so a tractor can drive through and spray Korean Natural Farming sprays to help suppress diseases.

Credit: Andrew Mefferd

"That's great," I said. "One of the things that appeals to me about no-till is that I know when we were trying to start a farm, we thought that we had to buy a tractor. I know a lot of people who have apprenticed around and wanted to farm. But the access to land, and then all the heavy equipment they have to buy, are big financial barriers."

"Yes, the heavy metal. The capitalization costs are so intense if you are going the tractor method," said Polly.

"Besides the fact that I'm a plant nerd. I'm not a machine guy. I see no-till as giving people a point of access to farm that might not otherwise have access to land or equipment," I said.

"Yes, it definitely does that. There are some other advantages to it as well. One is the space savings. If you're doing conventional row crops, you tend to have about two-thirds of the farm in uncultivated land, tractor pathways and so forth. This flips that. We have about two-thirds of the soil in crops, and only about a third of it just in tractor pathways. So we're able to get double the production from the same amount of space. The pathways are permanent. When we put down fertilizer, in this case it's compost or whatever other amendments we're using, we put them in the area where the plants are growing. We're not fertilizing the pathways. It makes irrigating a lot easier, too. The plants themselves shade the soil of a bed," said Polly.

"Because the space between rows is smaller, there is less exposed soil and therefore less evaporation. Also, the high organic matter soil holds more moisture meaning that we don't need to water as much. And with a high organic matter soil, when it does rain, the rain doesn't run off, it gets absorbed by the soil, which in turn means less watering," said Jay.

"Another thing; because we're not plowing, we can have a perennial crop next to an annual crop in any way, shape, or form we want, because we don't have to worry about ripping it up, or having to drive around it, or anything like that," said Polly.

"And we can grow season-long crops like peppers next to half-season crops like peas," said Jay.

Potato-Garlic/Winter Squash Rotation

"Potatoes are the disruptor. We move the potatoes around the farm on a three-year rotation program, and when we harvest the potatoes, because they're a big tuber underground, it ends up disrupting the beds," said Polly.

"Does that mean you get weeds that sprout up there?" I asked.

"We would get weeds if we didn't cover the soil right away. A strategy that we discovered was to plant garlic in the fall and then put compost down on top of the garlic. This way we are both mulching the garlic and covering up bare ground," said Jay.

"We do a mixed crop the year following potatoes of garlic and winter squash. As soon as these potatoes come out, the beds will be rebuilt with a potato hiller," said Polly.

"The soil that has gotten dumped into the pathways is put back into the beds, and then we plant garlic in alternate beds, and we cover it with compost. We cover the other beds with the compost, as well, and that smothers any weeds that are going to come up. The garlic will stay there all winter. The following spring, we put winter squash in the empty beds. The garlic comes out in early summer, which is right when the winter squash wants to take over. We just let the winter squash take over, and that's how we suppress the weeds [where the potatoes were], which is the only spot that really gets disrupted," said Polly.

A small field at Four Winds Farm.

Credit: Andrew Mefferd

"That's a great system," I said.

"Yeah, it works really well," said Polly.

The Armours use a tractor-drawn potato digger to speed up the potato harvest. It's the kind with a moving conveyor that lifts the potatoes, lets the soil fall through and deposits the potatoes on top of the bed.

"Yeah, and our potato digger, it doesn't really throw any dirt out. It's just lifting the soil, and it drops the potatoes behind," said Jenna.

"It doesn't do a lot of inversion," said Polly.

"It's not turning the soil. It just lifts it and drops it back down," said Jenna.

"There are some weed seeds that are brought up," said Polly.

Advice for Getting Started

"This system is scale neutral for the most part. So if you do it in one area, you don't have to do the whole farm. At a small scale like this, it's a four-acre market garden, basically. If someone is intrigued by the system but isn't really sure it would work for them, I would say the best advice is to pick the spot on the farm where you have the worst soil. Every farm has a spot that just doesn't do well, for one reason or another: it was abused in the past, it's really compacted, the soil is crummy, there's a lot of rocks, it's too wet, whatever," said Polly.

"Pick that spot, and try it, because first off, you've got the least amount of risk there. If it doesn't work for you, you've lost the production from the soil that doesn't do that well anyway. But secondly, if it does do well for you, that's the area that's going to show it most because with this system, you're not tilling, so you're not digging up rocks. You don't have to wait for the soil to dry out, so if it's a wet spot, you'll see the best benefit in that area, too."

"That's great advice. Would anybody need to know anything else? I was thinking, if I were going to do this, I might put down a tarp now. Exhaust the weed seed bank as much as possible, and then just lay out beds with a lot of compost in the spring on top of the ground that had been tarped," I said.

"We have seen people do something similar using cardboard, where they just put cardboard down and put the compost on top of the cardboard, and plant into the compost that same year. That's how they start. They don't even bother with the plowing phase," said Polly.

"We had a spot in one of our first years of doing this, where it just got away from us. We never got the compost down onto it, and it grew up in grass. Just a few beds grew up in really, really thick grass, crab-grass. Jay went in with the weedwhacker, and he just mowed it really close. Then we lay down newspaper, and then compost on top of the newspaper, and we planted carrots into the compost, and we walked away. We had beautiful carrots and nothing else. That really made it clear to us that this works well, and that was just newspaper."

Comfrey Border

"The comfrey does a great job of keeping the grass out," said Jenna.

I had noticed that all the fields on Four Winds Farm were bordered by some kind of plant that was loaded with little flowers. Turns out it was comfrey.

"Yeah, this is a barrier here," Polly gestured at the comfrey, "so that the rhizomes from the grass try and grow into the beautiful garden,

Kale behind the comfrey border.

Credit: Andrew Mefferd

which is lovely, loose, fertile soil. It loves it there, but it runs up against the massive comfrey roots. That's like the Berlin Wall. And the bees like it. And it defines the edge of the garden. Theoretically you can put your irrigation risers, headers in there, and they're not in the way where the vehicles are driving," said Polly.

"Another thing the comfrey does is create great habitat for garter snakes. One thing garter snakes eat are slugs. Needless to say, we do not have a slug problem," said Jay.

"It's a lot nicer to look at than a strip of plastic or gravel," I said.

"There you go. Other people like it besides us. Comfrey is great. Where you put it, that's where it stays. It doesn't migrate unless you were to try and rototill it, which is another reason not to rototill. We would have a comfrey farm if we rototilled at this point, and that would not end well," said Polly.

Another view of the kale field.

Credit: Andrew Mefferd

Same-Day Bed Turnover

"Have you bought a manure spreader? I know you said at first that you applied all the compost by hand," I asked.

"A manure spreader is designed to throw manure all over the place which is not what we are looking to do. Plus the wheel base of a manure spreader is too wide for the bed size we have," said Jay.

"Right now, it's a tractor bucket and a couple of people. So, Jay or Jenna drives, fills up the bucket and then just backs up slowly, and the people scoop it off. And then tidy it a bit. That's all they do," said Polly.

"Yes. That way it's just in the bed," said Jenna.

"Is there any bed prep after that, after you put compost on? Just plant into it?" I asked.

"That's pretty much it. The crew scuffle hoes, just to kill whatever small weed seedlings might be appearing. They scuffle just to break the surface a little bit, sometimes," said Polly.

"After things have been planted, and weeds have a chance to spread a little bit?" I ask.

"There are always some weeds. You're never 100% weed free, and so we use a variety of techniques," said Polly.

"There is always some organic residue on the beds from previous compost applications and a flame weeder would burn up that residue," said Jay.

"No, we don't flame weed in the beds. It's mostly just scuffling. We can pull a crop out, and just scuffle, and seed right into it. If it's really clumpy compost, and I'm doing carrots, or a fine seeded crop like that, then I might rake it out. But that's it," said Jenna.

"Yes, harvest and replant the same day," said Polly.

"Yes, that's what I've been doing. We've been pulling our onions out," said Jenna.

"Let's go look," said Polly.

"This was our onion block, and the same day, or the same hour that we're pulling the onions out, I'm just scuffle hoeing and seeding our next crop of spinach and beets there," said Jenna.

This is where the simplicity of the system really shines. Once the weed seed bank has been worked down enough and the tilth is good enough that there's no reason to till between successions of crops, the efficiency of being able to harvest and replant the same day kicks in.

The advantage of little time spent weeding and no time spent tilling really comes into perspective when you consider what happens on most farms once the onions are done being harvested; more jobs go on the to-do list—tilling the former onion area and remaking beds—before replanting.

The gain here goes beyond the time savings of eliminating two jobs. I know on a busy farm in the summer, jobs like retilling the onion area and remaking beds before replanting might find themselves below

Credit: Four Winds Farm

Tomatoes early on showing the straw and posts they use for their basket weave system of trellising.

other more pressing jobs like harvesting and delivering. If there is too much to do and retilling doesn't happen in a timely manner, the vacant onion bed will sit and grow weeds until such time as it can be tilled, generating more weed seeds and perpetuating the weed problem. The no-till system saves the opportunity cost associated with retilling between crops, and simplifies the farming system down to just what has to happen to get paid—planting and harvesting—and eliminates most everything else. In lean terms, it gets rid of the muda of tilling and weeding.

Low Weeding, No Thinning

"One of the great things about [the system] can be seen in how we grow parsnips and carrots. These are tiny little seeds that take forever to germinate, and the weeds almost always grow up first. Plus you have to plant really thickly, because you want to have a good stand. You don't want to have too many gaps, so you end up having to thin. What we do is, we don't do any of that stuff. We just plant the seeds, water them well, and walk away. This is what you get," Polly says, gesturing to a beautiful stand of carrots.

"With the carrots, we can use a lighter seeding rate, and then we don't have to go in and thin, which is another labor savings. And until you've thinned carrots all day long, you don't know what boredom means."

"Right now, with the height of the summer, we're spending our time picking and harvesting. I haven't weeded these carrots recently, and the last time I just took a scuffle hoe through. We just harvest them," said Jenna. "I did it in, I don't know, half an hour, 45 minutes."

"So this is a half-acre bed," said Polly.

"Some people are going to cry when they read that," I said.

"They're just not going to believe it. We know farmers who are very respected farmers, who insist that you have to till. Other than the potato digger, these fields, especially over there, have not been tilled in twenty-three years," said Polly.

"Here, Josh was five when we did this field. Is this Sasha's field?" Polly said, gesturing to the field behind us.

"Yes," said Jenna.

"Josh was five, and he's twenty now," said Polly.

"And it hasn't been tilled since?" I asked.

"No. After the first couple years of doing this, we sold our rototiller. The rototiller was great, but it was a crutch, because it's basically a giant eraser. When you have something that doesn't go right, or it gets away from you, you just come through with the rototiller and everything disappears. Until two weeks later, when it's all back with a fury," said Polly.

"Yes, rototilling is just such a paradigm. That's what most people do, right? Because I've been talking to people, telling them I'm working on this project. That I'm interested in no-till. I've encountered all of this skepticism like, 'Whatever they're doing, it's got to be more work than tilling.' Or just, 'Well, you've got to till at some point.' That's what really made me think, I've just got to go talk to people who are doing no-till, because seeing is believing," I said.

"Yeah, we've stopped trying to argue the point with the unconvincible. We just say, 'Hey, you want to see it? Come see it, we'll show it to you.' We're really glad you're here. Word has gotten out, in the last five years or so, suddenly there's a lot more interest. Not just with the crackpot farmers who are looking for something different, but from agricultural extension people, and researchers, and folk like that. It's the nuts like us who have been plugging along. Us, and Rodale, and so forth," said Polly.

"Some of these beds, a lot of this down here," Jenna says gesturing, "are full-season crops, like the carrots. But in the top section of this field, I'm getting three or four crops out of each bed in a season, by doing a lot of quick successions."

> We know farmers who are very respected farmers, who insist that you have to till. Other than the potato digger, these fields, especially over there, have not been tilled in twenty-three years.
>
> — POLLY ARMOUR

"Another cool thing we do is direct seed all of our onions," said Jenna. "We're not dealing with them in the greenhouse really early in the season, and spending all the labor of planting in flats. We just prep all the beds in the fall. We apply the compost, and then as soon as we can get in, we just scuffle hoe. Sometimes, I cut that out and just direct seed. Because the weed pressure is so low, the onions are able to get a start."

"Oh right, and you can get on the beds earlier because you don't have to wait for them to dry out enough to drive a tractor on them," I said. Most people in the northeast transplant onions for two reasons. For one, they need the plants to start growing before they can get on their fields with heavy equipment. And secondly, because onions are slow growers that don't compete well with weeds, they get a head start in the greenhouse without competition. I realize that when you can eliminate tillage and weeds, it opens up possibilities that are not available in most tillage systems.

"So, how much weeding would you do? Onions like these are direct seeded in the spring here. Have there been weeding passes through here, once or twice?" I asked.

"When they're really tiny, I do a quick hand weeding, just to grab any of the bigger things that are coming up, and then I did a quick scuffle hoe. Then we've done a couple hand weedings," said Jenna. "But often, the weeding is like, 'Oh, here is just one giant purslane plant.' And it goes really fast because the weeds come out really, really easily."

"Another thing you can do, you can mix different crops of different maturities in the same bed, because you don't have to think about planting it or tilling it after the crop comes out," said Polly.

Nearly weed-free beds of carrots and parsnips. Root crops like these can be planted tightly since they don't need any mechanical cultivation.

Credit: Andrew Mefferd

FRITH FARM IS IN SCARBOROUGH ON THE COAST OF Maine, just under two hours away from my farm in the center of the state. Since anytime I want to go almost anywhere else in the United States I have to head south past Daniel Mays' farm, I emailed him to see if I could invite myself by for a visit.

That's how I found myself on his farm on a drizzly October 26, 2017, taking shelter in one of his hoophouses while we talked and waited to see if the weather would clear up enough for us to walk around the field. Daniel grows organic vegetables on three of the farm's fourteen acres, and pasture-raises eggs, chicken, pork, and turkeys on the rest of the farm.

After starting the farm in 2010, Daniel developed his own no-till method similar to some of the occultation/compost mulch methods described elsewhere in this book. Some of what he has written on his website expresses the benefits and philosophy of no-till, and regenerative agriculture in general, very well. For instance, he "believes farmers should be stewards of the land, not miners of its resources, and that farms should be hubs of the community, not distant sources of its calories." He also believes that "economic sustainability need not be sacrificed, but rather can come directly from the union of environmental stewardship and community involvement."

Another thing I've heard echoed by a lot of other no-tillers is farming on a human scale: "The farm is sized for tools, practices and enterprises that celebrate the satisfaction and fulfillment of human work, and in return the work benefits from the increased care that this scale affords."

Looking out on his field from the door of a hoophouse, Daniel told me about his approach.

"I have an engineering background, so the layout is modular. There are sixteen plots, each of which has twelve 100-foot beds. They all rotate, so at this point it's basically the same crop plan each year rotated around the field," said Daniel.

DEEP COMPOST MULCH

FRITH FARM

Daniel Mays
Scarborough, Maine
Mixed vegetables, flowers
and herbs, poultry
*Occultation and organic
applied mulch/compost*

"That must be nice to not have to redo your rotation every year," I said.

"Yes, it is. Of the three high tunnels, those two had tomatoes, but we just ripped them out and now one has chickens and the other, winter greens. This one here has ginger still, since it's been such a warm fall. I'm eager to get the winter crops in, so we're planting the beds as the ginger comes out."

We look in a greenhouse where the residue from a tomato crop is still visible.

"Yep, so this had tomatoes in it until a couple days ago. We just pulled them out, raked off the leaf mulch, spread compost, and direct seeded into it. These greens will come online in mid-to-late February, and we'll harvest them right into May," said Daniel.

"So, the bed tops get a bunch of compost, and the pathways get leaves?" I asked.

"We mulch the whole tunnel with leaves before the tomatoes go in, and plant the tomato seedlings through the leaf mulch. When we pull the tomatoes out we rake the leaves into the paths and spread more compost on the beds. The raised beds give us extra space in the paths for the leaves," said Daniel.

Getting Started

"How did you get interested in no-till and come up with your system?" I asked.

"I had very little experience when I started out. I had volunteered on a handful of farms, but I hadn't spent more than half a season at a time farming. I read a handful of books on vegetable production and soil science, and all of them agreed that tillage is not good for the soil. So I thought, 'Why don't I not do this?' Though I started with a BCS walking tractor and rotary plow to break pasture and form the raised beds, after that, I saw little point in tilling more," said Daniel.

Farm owner Daniel Mays with partner Sarah Coburn.

Credit: Frith Farm

"The only reason I could see for it was to mix amendments and residues into the soil. But I rationalized that away: if you have enough organic matter, then the worms, soil critters, and plant roots do the mixing for you. It's a slow-motion biological mixing instead of a destructive mechanical one. And the soil responded right away.

"I also have pretty sandy soil, and I started at 3 percent organic matter. There's another farm down the road that has a similar soil type and I know they have had challenges keeping their organic matter up. I believe they till once or twice a season, or maybe even more, if you count cultivation. The soil can turn to beach pretty easily that way, whereas, you can't even see my soil anymore. It's so buried in organic matter from each year of mulching, composting, and cover cropping," said Daniel.

"You're making your own soil," I said.

"Yes, it's cool to see. Digging a profile now, you can kind of see back in time, how much soil has grown in the seven years I've been here," said Daniel.

Spreading compost, using a tractor to get it to the field and wheelbarrows to get it on the beds.

Credit: Frith Farm

"I'm imagining you didn't do this whole area in the first year," I said.

"No. I started with just under an acre and then added another plot or two each year. This year is the first year we didn't add any plots. We're at three acres total, including all the flower beds, perennial herbs, and high tunnels. This whole style of farming is not about starting with three acres. It's about starting small and building your market as you grow your farm, in harmony with each other. Then you can figure out what you should even be growing, based on your customers," said Daniel.

No-till outdoor cherry tomatoes.

Credit: Frith Farm

Cover Crops

"So, that plot with the sprinklers in it right there, is that a cover crop?" I asked.

"Yes. That's all cover crop. I'm experimenting with three different mixes, so there's four beds of rye and crimson clover, four beds of rye and mammoth red clover, and then four beds of peas and oats that will winter kill pretty soon. Next year, that whole plot will be solanaceous crops. I'm experimenting with the overwintered cover crops, and killing them right before the tomatoes go in, to see how that does, relative to peas and oats. From a management perspective peas and oats are easier, because they die over the winter, but then you lose that biological activity in the spring when the bed just sits there with nothing growing in it," said Daniel.

"So how are you going to kill the stuff that doesn't winter kill?" I asked.

"With the flail mower on the BCS [walking tractor] and then tarp it. I use black tarps. I joke that those are my 'organic Roundup.' Ideally, you could time it so that you flail mow the cover crop when it starts to go to seed, but the rye and clover will do that too late since I want to get tomatoes in there late May. So I think I'll be flail mowing them early May and then two or three weeks of tarping. Which I think will be enough to kill the rye and clover if it's sunny out, because the tarps not only block the light, but heat up a lot too. It's a nonaggressive solarization, as opposed to clear tarps. So, I think flail mowing and a few weeks of tarping will, hopefully, be enough, but it'll be an experiment," said Daniel.

"But that's how I do peas and oats. I do two and a half plots [out of 16] of peas and oats in the spring and then flail mow those late May, before they're really flowering, and that kills them pretty well and then I tarp them. It usually just takes about a week or week and a half of tarping."

"Do you have to do any weeding? There must be some little weeds that pop up," I asked.

Credit: Frith Farm

No-till beds after a thunderstorm on Frith Farm.

"Yes, we have some weeds. Basically, whatever we missed the year before, if anything goes to seed, which, this year, we've done pretty well at preventing. In theory there shouldn't be any other than that, but you always track some in with your feet from the pasture or some blow in. But we are able to maintain relatively weed-free beds. When we're done harvesting a bed, it's basically ready to go again. There might be a few bolted lettuce heads or something that we pull out by hand or flail mow if there are a lot of them, but otherwise we do no further bed prep," said Daniel.

I noticed that there was a lot of bok choy in the adjacent section of field.

"There was a mix-up with the seeding and we really overplanted bok choy this fall. Those beds we'll just flail mow and they'll be ready for the next crop," said Daniel.

"It's a bok choy cover crop," I said.

Credit: Frith Farm

Occultation with tarps.

"Yes. Exactly. What a cover crop," said Daniel.

"So, you take a crop out. And then, do you put more compost down and then seed a cover crop into that?" I asked.

"Yes, depending on the bed. I still treat the newer beds like that, building up the soil, but the older beds that have had that treatment for seven years are so fertile that they don't need more compost. For those I just plant directly into them. I try to always have a crop or cover crop growing to catch the nutrients that are already there," said Daniel.

The Biology in the Soil

"I am a firm believer in biology as the basis of soil health. I saw Will Brinton from Woods End Laboratories speak at MOFGA's Spring Growth Conference this year. What I took away from his message is that it's not the chemistry that we should worry about in the soil, it's the biology. What matters is the diversity and the quantity of your biology. Woods End offers tests that measure more than chemistry in a variety of ways. For instance, they can quantify biological activity in the soil by measuring the CO_2 released from the soil organisms in a sample," said Daniel.

"The idea of shifting from chemistry to biology, and the biological differences between the different types of organic matter, got me thinking. Compost is great, it's full of biology, but it's mostly humus and lacks the diversity of living and fresh organic matter that a cover crop provides.

"When you add compost to the soil, its biology is probably becoming food for whatever life is in the soil already. Whereas, when you grow a cover crop, the native, living organic matter that's being created is tailored specifically to your soil, and is alive and active, incorporated into the soil without tillage through root growth. If you're thinking biologically, cover crops are the best kind of organic matter.

"So that really got me thinking, because I've been a huge user of compost, as a mulch and slow-release fertilizer. Don't get me wrong, I'm still a compost junky, but now I'm cover cropping more. I think a crop after a good cover crop might do even better than a crop planted in four inches of compost."

"You can see a difference?" I asked.

"I'm pretty convinced there's a difference, because there's so much biology in the root system of that cover crop. When you kill that cover, all that biology is still active and looking for something to do. You plant your new crop in, and it's all right there. Compost on the surface is great, but it's not the same kind of active, grown-in-place biology that a cover crop provides," said Daniel.

"On Singing Frogs Farm that I was just visiting, they're really big on wanting to have roots in the soil all the time too. What they do is not dissimilar from what you're doing, as far as, they try to harvest a crop and leave the root system in the ground, and potentially even replant the bed the same day," I said.

CSA pick up on Frith Farm.

Credit: Frith Farm

"In fact, they have gotten to the same point that it sounds like you have, where they've got enough organic matter in their soil. They added a lot of compost early on and got to the point where their organic matter was high enough. So they stopped adding so much compost."

"Exactly. Plus, a cover crop is not really much management. You just seed it and wait, and let the organic matter accumulate on its own. I love watching the cover crop, like that plot of peas and oats over there that's thigh high. It gives me a lot of pleasure to look at it and imagine how well the next crop is going to do," said Daniel.

"It's hard to even tell that they're permanent beds at this point, the cover crop has grown up so much," I said.

Credit: Andrew Mefferd

A small part of the carrot harvest.

"Yeah. It's all kind of a jungle," said Daniel.

"So you have these permanent beds, and you're always walking in the pathways and always fertilizing the beds. And at the end of a season, when this bok choy cover crop comes out here, do you broadcast or do you drill a cover crop in?" I asked.

"We direct seed the cover crop with an Earthway seeder. It's a lot of walking back and forth, but it gives a real good stand, great germination, no wasted seed. I've tried broadcasting, and it's kind of hit or miss, based on how well you keep it irrigated," said Daniel.

"Yes, I've had the same issues," I said.

"I'll get two people, one on peas, one on oats, and they just stagger the lines. And it takes fifteen or twenty minutes with two people to do a whole plot of twelve beds. That's well worth the dense stands," said Daniel.

Field Management

The rain let up, so we started walking around, looking at Daniel's fall crops. We walked by some dense, healthy stands of carrots and rutabagas.

"So, I'm guessing you direct seeded the carrots. Did you direct seed the rutabagas?" I asked.

"The rutabagas, we transplanted. We do them every foot. For direct seeding, we pretty much just do salad greens, carrots, and radishes. Everything else is transplanted. Even beets are transplanted. It's a similar philosophy to direct seeding the cover crop. You just know you're going to get a perfect stand when you transplant it," said Daniel.

"This is where we had our garlic. Planting garlic and carrots back to back in a season is the most profitable plot on the farm. Those are both good return on the square foot."

"These carrots got planted sometime in July then, I'm guessing?" I asked.

"Yes, exactly, end of July. We harvest all the garlic, the next day come back and plant the carrots," said Daniel.

"It's nice to be able to flip them that quickly," I said.

"Yes, I've spent a long time looking at all the crop timings, finding the pairs that maximize the season. I have a color-coded crop plan I keep in the barn that is both the plan and the record, since we follow it exactly. It's helpful having that to go by, so the crew knows where everything is going. It takes the guesswork out of the busiest time of the year," said Daniel.

"And the quackgrass [at the edge of the field], I heard the farmers of Four Winds Farm talk about their border edging of comfrey. I love that idea, and thought about doing it, but it feels like a big commitment. If you ever want to move your beds, that comfrey's there forever.

Freshly mulched beds.

"That made me nervous, so instead we do a thick swath of leaf mulch around each plot. Over here I experimented. This path between plots was sod, but I just scooped it off and replaced it with wood chips. That worked out pretty well, so I might do that between other plots too. We need a path between some plots for garden carts or driving the tractor. When we spread compost, we line up three wheelbarrows, fill them with one dump of the tractor bucket and then walk them down the beds. That way the tractor doesn't ever drive in the fields."

"That makes sense," I said.

"That means I need access paths, otherwise I'm driving in figure eights around the farm instead of straight lines," said Daniel.

"How big are these beds? They look bigger than 30 inches," I asked.

"Yes. They're 60 inches wide, on center," said Daniel. "That makes the bed itself about 42 inches. I like the idea of the 30-inch width, because most of the implements on a BCS can cover that in a single pass. But if you do the math, you lose over a third of your growing space to paths. I have long legs, so I can still step over these beds easily enough. That's why I went with 60 inches on center when I started. If I were starting from scratch, it would be a tough call, because the single pass is a pretty good selling point," said Daniel.

Compost

"I buy two different types of compost. This pile is purely composted leaves, with no other nitrogen source. I use that on most of my beds, at this point, that don't need any more nutrients. It's basically stabilized organic matter without extra phosphorus and nitrogen. I use it as a mulch in the fall that we can just plant right into the next spring. We use the regular fertile compost for things that need more fertility. We'll spread that before we plant garlic, then we don't fertilize the garlic the whole next season," said Daniel.

Cover cropped no-till beds.

Credit: Andrew Mefferd

"So I imagine your soil organic matter has come up, if you started out at 3 percent," I said.

"Yes, it's over 6 percent now, which for super sandy soil is pretty good. This style of farming doesn't really fit into the paradigm of soil testing, because if I test the top four inches, it's almost 100 percent organic matter. If I test the soil eight inches down, organic matter is probably still pretty low. So I guess I'm testing the average of the top ten inches, but based on how deep I plunge the sampler, it's going to be a little different each time. Those tests are all based on the idea of tilled, homogenous soil. My soil profile is more like a forest floor," said Daniel.

"Was it hard for you to establish this system? Did you have to fight weeds that first year to kind of get over the hump?" I asked.

"Yes. I've definitely had to learn the hard way about weeds and not getting too big so fast that you can't stay on top of them. I've let some weeds go to seed some years when I didn't have adequate labor. You really pay the next couple years, because you don't get to just till it under.

The mobile chicken coop on Frith Farm.

Credit: Frith Farm

It's all right there on the surface. This year we've pretty much caught up with that and stayed on top of the weeds, so I'm hopeful for minimal weeding next year," said Daniel.

"We spread a yard and a half of compost per bed. That's well over a thousand pounds per bed. I go through a lot of compost. My solution for spreading it is a hybrid tractor-human model. The wheelbarrow is an amazing human-scale tool, easy to handle for extended periods except for the loading of it. That's where all the work is. So the tractor does the loading right at the head of the bed," said Daniel.

"Once loaded, the wheelbarrows get pushed down the bed by hand, and dumped precisely where it's needed. It works pretty well. Four of us can spread a whole plot with fifteen cubic yards of compost in about twenty minutes. One on the tractor and three people pushing wheelbarrows. It's actually rather fun. It's a team effort, instead of just me loading a manure spreader, dragging it around and compacting the soil. It's almost a festive thing. Everyone's working together to keep up with the tractor," said Daniel.

"I feel like there is the ideal system once it's established, and then there are the systems of getting it there, which often involve tilling. But I'm looking at these beds that we just sheet mulched and planted right into. If you spread enough material that it smothers the crop, or lay down cardboard and then compost, you can avoid even the initial tillage to establish beds. It takes a lot of organic material, but ultimately it pays off as it feeds your soil for many future crops," said Daniel.

HEDDA BRORSTROM GROWS A DENSELY-PLANTED acre of cut flowers at Full Bloom Flower Farm and Floral Design in Sebastopol, California. She was supposed to be the first of a few visits to no-till farms in and around Sebastopol in October 2017, which turned out to be really bad timing. The night before I was supposed to visit her farm, historically bad wildfires broke out in Sonoma County, which meant she was busy helping friends out for the next few days and I never made it by her farm.

So I was very glad to catch up with her over the phone early the following spring, and hear about how the biological activity on the former worm farm is helping to break down crop residues instead of tillage.

FULL BLOOM FLOWER FARM & FLORAL DESIGN

Hedda Brorstrom
Sebastopol, California
Cut flowers and floral design
Compost mulch, occultation

Getting Started

"How did you get started with no-till?" I asked.

"I started out as a gardener. I went to UC Berkeley and I studied agriculture, but more the politics of agriculture there. And I knew at that point that I wanted to become a garden teacher. So I was a garden teacher for six years in San Francisco, and managed an acre garden with 450 kids. And I didn't have any machinery there, so tilling wasn't even part of my thought process when it came to farming. And then when I decided that I wanted to learn more about farming I went to the UC Santa Cruz apprenticeship program and did that six-month training there. They have three different sites that you learn on. One is a twenty-acre tractor-scale farm, then there are two gardens," said Hedda.

"Both of those gardens implement French intensive methods [for maintaining] everything. That seemed like more my speed, and I had already fallen in love with flowers, and realized that I could farm a lot less land and have a flower farm at my parents' house in Sonoma County. So I leased land from them, about an acre, and figured I could do flowers, since not that many people were doing flowers in Sonoma County at that time, and make it a viable business. I'm going into my sixth season

Credit: Dawn Heumann

Hedda Brorstrom in the densely planted, closely spaced beds of Full Bloom Flower Farm.

now. The first season, I had a tractor come and rototill the field. And half of the field I [tilled with] a walk-behind tractor the next year, but half of it I had already started no-till farming, without really calling it that.

"It was all annuals at that point. Then about the second season in with using a rototiller, I just didn't want to anymore. My soil was really, really sandy, and it didn't seem like I even needed to. And I had done so much sheet mulching in the past, that was what seemed instinctual to me. So I did that for about three seasons on half of the farm. And then for the last three years, the whole acre has been no-till. I could transition because the beds where I had been doing no-till had a lot less weeds.

"The whole in and out of flowers is a bit different than vegetables. Another part of my farm, and making it something manageable to be a one-woman show, is that it's a couple of businesses. I grow cut flowers, a lot of them, but I also have design work; I'm a florist on the side, and there are a lot of weddings.

"So in order to have that many cut flowers and be selling to grocery stores and marts, and designing for eight restaurants in town, and all these different accounts, my game plan from the get-go was always to start investing in perennials. So slowly I've been adding more and more perennials to the field, which obviously are going to be in a no-till system.

"One part of my farm that's unique is that my spacing is about the closest I think I've ever seen on any flower farm. I'm trying to squeeze as many plants into as little space as possible. About half of my bigger field at this point is all perennials. For a long time, I had been really attached to the idea that I needed to do a lot of cover cropping. I just happened to start farming in a drought, so it wasn't until last year that I had really experienced what it was like to farm with rain. I just kept trying to grow cover crops, and it wasn't working, because I don't have anything but drip irrigation. I don't have any way to water in the off-season, and the flowers go all the way until frost."

The Method

"I have kind of given up on cover cropping. I don't cover crop, and the way my system works is, I start plants in October. I have about five hundred bed feet of spring crops that really produce a lot. Right now it's anemones and ranunculus. Those were uncovered in the past, this year I did low tunnels on them. So there are only five production rows through the wintertime," said Hedda.

"And then in the rest of the beds I cut the plants, so I leave the roots in every year, and just cut the tops. Especially sunflowers and amaranths, things that have really big roots; I don't make the effort to go through and pull any of them out, so I'm just chopping down material, and either laying it on top of the field or removing it if it seems like there is too much material.

"But most of the time I'm just throwing [the plant matter from the previous crop] down, and then for the winter pulling landscape fabric over sections. And then when spring comes, using one bed at a time and gradually pulling them off. So this week [beginning of April] we pulled all the landscape fabric off the small field.

"We use a lot of compost. I think one of the problems that some of the other vegetable farmers are facing now that have been doing no-till, is that especially in Sonoma County, we don't have great compost anymore. And people were adding a lot of compost with a lot of animal in it, cow manure, and the phosphorus levels were getting really high in some people's fields. So I'm switching off a lot between different [composts]. And because I'm a flower farm, I'm not having to be as cautious with the application of manures. Last year I did it way too thick. I did

I'm consistently planting, I would say a month and a half before other flower farms, because of being a no-till farm, and being able to get into the ground extra early.

—HEDDA BRORSTROM

Credit: Dawn Heumann

Credit: Dawn Heumann

Some of Hedda's design work.

this incredibly thick layer of duck compost, almost eight inches thick."

"Wow," I said.

"Yes. I was piling it on my beds, and I didn't even broadfork anything. We used to be a worm farm, and nothing from last year exists anymore [it all decomposed]. The worms have just turned it. So that was pretty fun to see how much the biological activity is moving through the plants. There's turnover from spring crops. I'll do a plant-out right now of some early spring crops. And when I transition from those to late fall crops, I often will either just cut the plant out and remove it, and then direct seed in the hole, and add a little more compost or just transplant and add a little compost as I'm transitioning the plants and reusing the beds," said Hedda.

"You mean you're just taking the biomass, the top growth of the plant, off and just making a hole and sticking a seed in?" I asked.

"I meant, I'm probably harvesting a spring crop, and taking the biomass off and leaving the roots, and then just sticking a seed at that same spot that caused the drip; I'm often just planting by my drip lines," said Hedda.

Mulches

"Compost application this year is going on a little bit less thick. We still like four inches. I'm still doing this initial [application], because those beds haven't been no-till for as long," said Hedda.

"I was already no-till farming, and then when I found out about Singing Frogs Farm (see interview p. 275) I thought, 'Oh, I didn't even realize that [no-till is] a thing that other people are doing. That's so cool to have them as my neighbor.' And then I've had a lot of the crew members from Singing Frogs come to my farm just to check out my system too.

"And it seems like we do a lot of very similar things. This year was the first year that I had bought a silage tarp for one of the fields, and did silage tarp instead of landscape fabric. Which didn't make me feel great. The plastic is pretty stinky. Part of me thinks, 'I feel like this is really good to not be tilling.' Another part of me feels like I own way too much plastic.

"So I've got this huge silage tarp that doesn't let the water in, which is kind of a good thing. That part of the field was too wet anyhow. And I'm consistently planting, I would say a month and a half before other flower farms, because of being a no-till farm, and being able to get into the ground extra early."

"That's a big advantage," I said.

"Woven landscape fabric, as the solution to small-scale farmers not having to weed as much, works well where it's a little colder. Here we can't grow as many types of crops in it because of the heat. Last year we got up to 112 [degrees Fahrenheit/44 Celsius] in my field on one of those days. But normally it's more like 102 [degrees F/39 C]," said Hedda.

"As much as I can, I've been trying to use Weed Guard Plus [brown paper mulch], and then using straw on top of that, or organic alfalfa when I can find that as well. But it's not as easy as rolling something out and then rolling it back up and moving it. I've been using the straw on those parts of the field that have been no-till for five years now. And I decided it just makes me feel better about that whole zone. If I had access to wood chips, that would be awesome too for all the perennials.

Dahlias being planted in crates to keep the gophers out.

Credit: Full Bloom Flower Farm

Credit: Full Bloom Flower Farm

Later in the planting process the tubers are covered with soil and straw.

"Another huge thing that I found with no-till farming in our area that's a big pain and a drawback, is the fact that the gopher population is harder to find. If you till a bed in the springtime, you can really go out there and trap pretty easily. But I'm planting plants so quickly and in succession, that it's harder to find them. And then when the landscape fabric is down, they're just building these amazing tunnels for months at a time. So that's one of the other drawbacks: the plastic and the gopher issues."

"But in general, it's been amazing not to have to use any machinery. Have you ever heard of Bare Mountain Farm?"

"Yes, I visited them last October," I said (see interview p. 123).

"They have really good explanations on practical farm things, with a great YouTube channel. And it seems like more people have discovered no-till in the flower community because of them," said Hedda.

"And it just so happens that one of my best friends here owns Red H Farm, she's a vegetable farmer. She does about an acre and a half or so. And she's a no-till farmer too. So without having to try too hard to find other people to talk to about it, it was just part of this community. And that was even before we knew about Singing Frogs. Since then I think I've convinced at least four other little farms around here to at least try doing one of their fields in a silage tarp or landscape fabric.

"I started this group called the North Bay Flower Collective. There were five people originally, and now we're up to fifty people, just in Sonoma County. The flower scene has really bloomed here. We have these monthly farm tours, we just go and check out each other's farms and learn. I would say that most of the flower farms around here have at least one or two fields that they're experimenting doing no-till.

"It's been interesting to see it kind of sweep into Sonoma County. This winter, all those people were buying silage tarps for no-till farming. And now it seems like a lot of the farms are adopting it, especially with finally having rain, and being able to start farming a good month before they used to."

Planting

"When you said that you space things lot more tightly than most people do, did you mean in-row spacing, or you crowd the beds together, or both?" I asked.

"Both. I'm putting most of my plants at six to eight inches apart, and then doing either five or six rows per bed. So I'm fitting in a lot of plants. And with flowers, that makes them extra tall, which is always a goal within flower farming. I've added four hundred feet of roses, and they take up a lot of space and don't produce as much, you've got to space them out so far apart. Everything else gets treated like every square inch is critical. So things are really, really tight," said Hedda.

"When I started out originally, I wasn't thinking I'd have a lot of helpers. Now I definitely have to have help on harvest days. Even now [early spring], when I only have five beds in production, it's a six-hour harvest. So my rows are a little too narrow for some people to get down some of them. It's almost a maze. But in a lot of ways, it really helps with the flowers, because I don't use any netting.

"And so they stay up, they stay more upright than other farms because there are just so many of them. There's not a lot of space for wind. But that's something not everybody could do. My field is really hot and dry in the summer, so I don't often get any powdery mildew or anything

The beds at Full Bloom Flower Farm.

Credit: Full Bloom Flower Farm

on my zinnias. So I'm able to put my plants at this crazy tight spacing that other people couldn't get away with. Some things go at nine inches, zinnias go at nine inches. But sunflowers are going in at six inches, so you can really put thousands of plants in a really small space."

"I think that makes a lot of sense. It's almost like approaching your field like a greenhouse. You know how everybody tries to crowd as many plants as possible into a greenhouse? So if you're using that planter's paper, you must have problems with weeds coming up through your beds?" I asked.

"I still have weed pressure and issues in my field. But the sections where I've been doing no-till the longest, they definitely have fewer weeds. I did one crazy experiment, that is no-till, but it's not really any normal farming. I have a crazy plant-out of four hundred bed feet of dahlias on top of landscape fabric in crates. I did that because I was starting to get a grass that looks like crabgrass, and it was just driving me insane. So I thought, 'You know, I'm going to cover this for two years and do my dahlias in crates because the gophers are such an issue [eating the dahlia tubers],'" said Hedda.

"Then storage has been an issue. And I've been able to save every tuber that way. Even leaving them out in the rain. It's kind of a crazy system that a lot of people are copying now. And I tell them, 'I don't actually know about this, I'm just experimenting.' Before that I was doing these trenches with gopher wire, but they rusted out. This is season three of doing the crates and the dahlias, and it's crazy expensive. But that was probably my biggest weed issue, that grass, and now it hasn't been too bad, but definitely not weed free. Especially since it is primarily just me on the farm, and I don't really allocate much time to weeding."

Types of Compost and How Much to Use

"When I did the crazy thick compost layer last season, I don't think I weeded once. So that's kind of my M.O. on doing the thick compost. But at the same time, it also killed a lot of plants because it had been just too thick and they weren't getting water properly. It was rice hulls mixed with duck poop, and it would keep it dry," said Hedda.

"Some people I've talked to who are doing no-till, they make a really high carbon compost, so they can apply a lot of it without, like you said, spiking their phosphorus or other nutrients," I said.

"Which is funny, because one of the main facilities that I was using for compost for years, was really, really high [carbon], and it looks just like wood chips. So I switched from using that for years, into using the duck poop last year, and now I'm back to a different compost that's municipal waste plus grape seed extract from the grape industry. And I also used cow this year too. So I've been doing 30 or 40 yards for a little bit less than an acre per year," said Hedda.

"Yes, that's a lot of compost," I said.

"It's a lot of compost, especially since I don't have a tractor, to apply with a bucket. Because one of the neighboring farms, they're doing no-till, but they still bought a tractor just so that they could use the tractor bucket to lay the compost out. So it's not a tractorless system for some people. My field's just so small, I can't bring a tractor in anymore, because there are so many perennials," said Hedda.

Flower beds with a bouquet.

"So are you just wheelbarrowing all that compost around?" I asked.

"Yes," said Hedda.

"Well that's a workout," I said.

"Yes. Spring is real," said Hedda.

A Very Diverse Operation

"You said you had a lot of perennials and the dahlias. Are there particular crops you specialize in, or are you pretty diversified in what you grow?" I asked.

"I'm pretty diversified, but I've gotten more specialized over the years, because my main

Credit: Full Bloom Flower Farm

moneymaker is weddings. The way I think about it now is, I have ranunculus and anemone season, then I have rose season; they're all heirloom David Austin varieties, and then it's dahlia season and lisianthus season. Those are the focal flowers, and the bigger sections are about five hundred feet of each one of those things. And then about two hundred other little things. Sometimes even just three feet of a special grass for boutonnieres or something," said Hedda.

"So it's really diversified, but I'm getting into larger plant-outs. And then I sell a lot to the fifty other florists that are in our group. So a big part of my sales goes to florists now. It used to be to the grocery stores, and I used to grow more market-style things, more sunflowers and zinnias, but I'm reducing all those numbers just because of the fact that we're such a wedding destination, and that's where the money is," said Hedda.

"I'm also adding other things in that I used to grow, because I was a food grower before. So this year I'm doing strawberries again, because the florists love them green for arrangements. So we're definitely Sonoma County. Sonoma County flower growers are specializing in unique and boutique varieties; plants like Stainless Steel roses, Distant Drum roses, sea oats, clematis, scoop scabiosas, and other wedding items. Santa Cruz farms tend to do high volume and brighter colors while we are more focused on main wedding focals and then the odds and ends that lend a rustic touch to design. Hops, lilacs, cherry tomatoes, and anything odd or unusual like mushrooms gets bought up very quickly at our local flower mart."

Differences Between Growing Flowers and Vegetables

"A lot of no-till veggie growers are taking advantage of the fact that they don't have to till from one crop to the next. They're getting one crop out and then replanting the beds to another crop the same day, or at least very quickly. But that speed of succession doesn't really exist in flowers. There's not a salad mix of the flower world. How many times are you getting to use the beds over the course of the season?" I asked.

"A lot of the beds are just one crop per season, but some of them are two. Rarely I would say that it's three. And because I'm trying to do more and more perennials, there's a big portion that stay planted for years. But then, I went into my fifth year last year, and realized, 'Oh yeah, there are a lot of things that have to be replaced after four years.' And that was really apparent in the springtime. Just that I needed to dig things up and separate things out like irises, and these bigger plant-outs of Veronica and the perennials. It's not like vegetables where you're constantly planting all the time. I think my last seeding date is mid-July," said Hedda.

"So it's very different than with vegetables. I start seeding January 28th most years. So late January through July. A lot of the other spring stuff is fast. Like stock and forget-me-nots, larkspur, sweet peas. I'll plant on top of those when they're done, and do another vining crop like love in a puff or hyacinth bean on top of the sweet peas. So it's mostly just the spring crops that get a second round on top of them.

Hedda in the field in the fifth year of the farm.

And some of the grasses too, I do a lot of different grasses, like ruby silk and frosted explosion. And some of the scabiosa too, like *Scabiosa stellata*, I try and get three rounds of that in. And as things end, I'm just going in and plugging in zinnia seeds, direct seeding, as the spring stuff is finishing, and I don't have replacement plants."

"Do you have any advice for people who want to get started with no-till?" I asked.

"My advice would just be to try a section out. It's so satisfying to pull back whatever medium you're using to cover, and have a bed pretty much ready to go, and see the biology that is so alive underneath it. Time and time again, that does not get old. And I think people inherently know, whether or not they're soil scientists, that when they pull that back and there are all those worms and all the insects and all the microbes, and all the fungus, that it's a really good starting point. And that they didn't really have to do anything, and it was easy to just move something versus starting an engine up," said Hedda.

"So that would be my advice, to try out a hundred feet and see how it goes. And don't expect the promises of no weeds to be the motivator as much as seeing the worms as the fact that you've got these little guys underneath helping you out."

"Yes, that's great advice. Just try a bed. That's another thing I like about no-till, it's not like you have to invest in a whole lot of equipment," I said.

"Yes, and it's not complicated either. It comes instinctively to people once they've done it a couple of times. And just to go to each other's farms, that's always my biggest advice. It's awesome what we can do and watch online, there's so much information. But going to another farm, that's priceless. Whether it's a farm that's been in operation for six months or sixty years, you're always going to learn something by going to somebody's farm. I'm a big advocate of, if you're interested in no-till, go and visit a no-till farm and ask questions," said Hedda.

IT WAS STILL SMOKY FROM WILDFIRES WHEN I DROVE out to Shanon and Michael Whamond's Hillview Farms in Lincoln, California, northeast of Sacramento. On a clear day you could see Sacramento. Tucked in among ranch homes, I found Shanon and Michael's beautiful new farm.

Shanon and Michael Whamond
Lincoln and Auburn, California
Mixed vegetables
Occultation, compost mulch

The farm is ten acres of sloping land surrounded with deer fencing, a walk-in cooler with a colorful mural on the side, and a barn. After outgrowing a nearby piece of land, they had only been farming here for just over a full season when I visited. Shanon and Michael embody one of the biggest potentials I can see for no-till: the ability to start a commercial farm with almost no land and very little equipment, get a business going, then scale it up as necessary.

When Michael and Shanon wanted to start a farm, they didn't have any land. Michael's dad suggested they use some of the land around his home in Auburn, also northeast of Sacramento, and rent the unused pasture next door to get started. So that's what they did. Five seasons later, they needed more land to grow on, and were able to buy the new farm in Lincoln. Not only does the property already have the infrastructure for farming, but also they likely saved it from development.

"There was a CSA farm on this property for about four years before us. They put in immaculate infrastructure that only a farmer could truly appreciate—deer fencing, barn, walk-in coolers, pack station, and irrigation. We were grateful to maintain the farm's integrity and continue to farm the property," said Shanon.

"Right now we're in transition. We started on a three-acre parcel. At first we thought, 'This is so much land,' but then we realized we needed more land. So we leased land from our neighbors out in Auburn and once we filled that space up, then we got this beautiful ten-acre parcel."

Getting Started

"We have no intentions of farming the whole ten acres. We definitely want to do the intensive small-farm model. Maybe cultivating about two acres would be our max. When we first started farming, we were pretty naive. We were definitely swayed by the whole romantic aspect of it, thinking we would just frolic in fields all day, but we learned pretty fast how demanding farming is. We didn't really know at what capacity we wanted to farm, so when we first started, we looked at probably ten different farms just trying to find a model that made sense for us," said Shanon.

"We visited thirty to forty farms, because we had no agriculture experience. So we just jumped into the deep end of the pool. When we started farming on that three-acre parcel, it started off as a half-acre. I realized very quickly that we weren't going to make enough money. Since it was an urban area, the plot sizes for our rows were about fifteen feet to two hundred and fifty feet, and everywhere in between," said Michael.

Greens on Hillview Farms.

Credit: Hillview Farms

"When we got a tractor we couldn't really cultivate those areas, there aren't any turnarounds in a space like that. And then we would go on farm tours. I would see these hundred-acre farms, and I thought, maybe we can't do this," said Michael. "This is not viable. And I started researching soil health and the one thing that kept on popping up was no-till. You know, tilling is destructive to all the organisms and everything in the ground. So we took a field trip out to Singing Frogs Farm and that —"

"That was just kind of a light for us," said Shanon.

"Yeah, it was like holy smokes, and just Paul and Elizabeth in general, especially the way they talk, they're very convincing," said Michael.

"One thing that really impressed me when I went there, it was early March. And I went to three other farms on the tour before I got to theirs. And all the other farms were just starting to mow down cover crop or starting to till and getting ready for the spring and summer plantings. And they had three acres of full production in the ground. The biggest thing we were concerned about was the profitability of small-scale farming. And when I saw that, I thought, we can do this on a different level and use less machinery to get where we want to be. So that was our intro to no-till. And then we came back."

"And we tried to set up the system that Paul and Elizabeth have, but change it so it worked on our acreage, because we were in Auburn with heavy clay soil. We knew it wasn't going to be a turnkey kind of thing so we tried out a little bit of everything. Some things worked, lots of things failed. And then we went from there. It's an exciting time for small farmers right now because no-till is exciting, but there are also all these new tools that are coming out that are made just for small farmers. So we just jumped on the bandwagon and were like, 'Let's do this.' It's not like we're like signed up for any one style of farming. We want to farm the way that makes sense on our land and at our capacity. So we're no-till, but we don't put ourselves in a box. For example we do use a Tilther to prep beds," said Shanon.

"I don't know where that falls in line," said Michael.

"To my mind it's still no-till. I think a big part of it is not inverting the soil layers, and that Tilther's only working the top inch or so. It's not much more soil disturbance than what you'd get scuffle hoeing," I said.

"It just depends on who you talk to. Our introduction to no-till was Singing Frogs Farm and that is no-till. There is no machinery on there—" said Michael.

Credit: Hillview Farms

A carrot pulled from the soil at Hillview Farms shows the worms that are proliferating under no-till.

"They're awesome," said Shanon. "Our first year we started actually farming, the no-till system really worked for us because it was a lot of hand labor, so we were able to manage things just by us two. So we started growing things and then we needed to sell it somewhere. That's when we started our farm stand, which has been really awesome. Our original satellite location is right in the middle of a residential area, right off the main road. Hillview Acres is the name of the association that we're a part of, so Hillview Farms is an ode to that. Which is funny because it's not on a hill, it's in a valley."

"And our new place is on a hill so now we're set," said Michael.

"So we started the farm stand and it was really great. Started out with just our friends and family. We would convince them to come out and support us. And then from there it grew into this awesome community thing. And we just started small. We got into our local farmers market," said Shanon.

"We actually didn't get in at first, but we had planted all these starts. So we asked them if they needed nursery starts. And they said yes. So that was our foot in the door. And also, no-till allowed us to farm for four seasons. Not at a full capacity, but our off-season yields gave us a foot in the door in the farmers markets, which was really good."

"When you're talking about cash flow and things like that for the second year of farming, we actually wouldn't have been farming if we didn't switch most of our plots to no-till and keep on consistently planting in the ground through the winter. That gave us our cash flow for the springtime because we were farming during the wintertime," said Michael.

"We wouldn't have had the capital to start up and buy seeds again if we stopped farming. So we just said okay, good, we did six months. Let's try again. And we chose not to do a CSA model just because of our knowledge. It was too daunting of a task. For people to give us that much money where we might fail. That was really key to get us started and to start small.

"We started off with just ten thousand square feet at the Auburn farm. Eventually it got to be one and a half acres. And we quickly real-

ized that for us to be more efficient, we needed
to systematize and do 100-foot rows. All in a
line, all together, and then be able to grow from
there instead of jumping around in people's lots.
Because in our second year, 75 percent of the
farm was on leased land. And it was really scary
to think about if we lost that land. So this property in Lincoln came up and we've been here for
one full season."

Becoming More Efficient

"It's funny, coming from no-till, I just tilled
that," said Michael, gesturing to a freshly tilled
area they're going to start growing in next year.
"That's how we're setting up beds."

Credit: Hillview Farms

Harvesting lettuce from
beds that have been
covered with landscape
fabric to suppress Johnson
grass and other perennial
weeds.

"To set up the initial plot, we'll grow a cover crop. Mow it down,
till it under, and set up permanent beds. Our first plot right here hasn't
been tilled since we started. So it'll be about a year and a half," said
Michael.

"So, till, plant a cover crop, till the cover crop in, is that to get some
organic matter in the soil to start with?" I asked.

"It's to shape the beds. We want to do a raised permanent bed,"
said Michael. "They don't have to be super high. But the easiest and
quickest way to make permanent beds for us, has to do with the type of
machinery that we have. Especially with our heavy clay soil, once it's in
place you don't really move it around. It's very dense. What I see people
doing, everyone is using a [walk-behind tractor] with a rotary plow to
make raised beds. We got a [riding] tractor in the beginning, so we don't
have a [walking tractor].

"So we use a tractor to move compost around. We need that tractor.
My first eight months of farming, we were doing heavy composted beds
and I was doing wheelbarrows. Wheeling them around three acres. And
holy smokes, I would not have farmed one more month. It's just too
labor intensive."

"Yeah, that'll age your back quickly," I said.

"So we have a tiller, I just made this bed shaper. We did these [first beds] by hand, and it's just not ideal. The one thing we found with no-till is that it requires a lot of labor. For us, starting off, we're learning how to grow the vegetables, we're finding markets to sell them. Now as a business, we're just starting to get into the management of people. And that's a whole different skill. That's very tricky," said Michael.

"So we would be working on the systems and implementing as we go. We're trying to use some of the technology that bigger market growers are using. And the efficiencies that basically take out the labor, with small machinery. So that quick-cut greens harvester was a huge help."

"I love that thing," said Shanon.

"The Tilther helps prep beds. Beforehand, we weren't able to get very good seed-to-soil contact when we just laid compost down.

The view on Hillview Farms the day the author visited. The amber hue of the light is from smoke due to wildfires burning in the region.

Credit: Andrew Mefferd

Tilthing helps, and then using a Jang seeder helps. We used to do everything by hand transplants. So we did plugs. Like this arugula right here. We transplanted nursery plugs," said Michael.

"We do it also to keep up with weed pressure so they're established before they go into the beds," said Michael. "It helps out. Now the biggest crunch on it is our labor. Though it goes pretty fast. I can do a whole hundred-foot row in about thirty-five minutes.

"And then we're doing some carrot trials and root crop trials to expand our variety. We haven't grown carrots in four years. We did it our first year. We never did it again because of the heavy clay soil at the other farm. There was no soil. When we did the soil analysis, it was a class seven soil. And I was like, 'What the heck's a class seven?' We only thought there were four classes. We didn't even know there was a class seven. So basically, when we first started off, the rototiller was bouncing off the ground because of the amount of rocks," said Michael.

"It was so compacted," said Shannon.

"So the amount of compost that we added to the soil to amend it was a lot. The soil [at the new site] is already [better], there's about two feet of topsoil. And for us, this is a heavy clay that we're in love with because we're used to it. So it works for us and it works for our system."

"It grows beautiful veggies," said Shanon.

"But you can see we're falling into that greens pattern. Marketing a lot of greens and things like that. We're doing lots of tomatoes," said Michael.

"We're excited for the opportunity to really just settle in for the first time since we've started. And just fully saturate our markets and become as efficient as possible with what's in front of us at this present moment. And then scale out from there," said Shanon.

"So it is exciting. And the crops here are tremendously healthy," said Michael.

"Everything looks great," I said.

"This looked completely different two weeks ago. After the intensity of this summer and all those crazy hundred-degree days," said Shanon. "We were so eager to pull our summer veggies out."

Weed Management

"One of the biggest differences between us and Singing Frogs is weed management. And when you go to Paul and Elizabeth's farm, one thing they say is they don't weed. Did they say that?" asked Michael.

"Yes. In fact, I asked them about their weeding and they said to ask you guys about that," I said.

"That's funny," said Shanon.

"So when we started the farm, over at the Auburn plot, it hadn't been used in any agricultural setting. It was an urban area, with native grasses, and when we set up those no-till plots over there, we did absolutely zero weeding," said Michael.

"Well the weed seed bank wasn't that intense," said Shanon. "I wouldn't say we did zero weeding. We did some."

"We just used compost to smother the weeds and we probably spent thirty minutes the whole year. I mean, we would be walking by and were like, oh okay, there's a weed," said Michael.

"So there did you rototill, drop compost, and just plant into the compost?" I asked.

"Yes. We used compost as a mulch. Basically what they suggested is about two to four inches of compost on top to establish the bed," said Michael.

"Is that what you did here on the new farm?" I asked.

"Yes. Not as much compost here because the topsoil is so different. There's already a really good amount of top soil," Shanon said.

"Some of these beds, we did about two inches. We never went to four inches. We didn't put that much compost on top because I felt like it didn't need it. And the biggest difference between weed pressure at the Auburn farm and here in Lincoln is that we have rhizomatous weeds," Michael said. "We have Johnson grass, bindweed, and Bermuda grass and holy smokes, when you lay the compost down as a mulch, they don't care. That's not a mulch for them. It's a feeding source to sprout up."

"And it comes up right through the compost?" I asked.

"No doubt. You can't really tell, because we did heavy occultation and covering and just matting and tarps, matting it down for about a year," said Michael. "And it works, but those weeds—"

"Man, they'll come up. If you don't stay on top of it, Mr. Johnson's gonna take over your farm real fast," said Shanon.

I noticed they had some beds that were completely covered up. "Is that why the beds over here have landscape fabric on them? Is that to keep the Johnson grass down?" I asked.

"Definitely," said Shanon.

"Our biggest Johnson grass plot was over here in the corner. It's unbelievable. There's Bermuda too. So you got Bermuda, Johnson, bindweed…" said Michael.

Digging in.

"When Michael said we did the occultation, the tarping, we used those big silage tarps. They're white on one side, black on the other side. So we'll take those and spread them over this whole bed section, put sandbags down, hope that the wind doesn't pick up, and cover the whole bed," Shanon said.

"So from here over, we had our early-season tomatoes. And in the fall of last year we tarped it. It was tarped from November to March. For three to four months," said Michael. "We pulled off the tarps and there was already Johnson grass emerged under the tarps. It was yellow, but it was still sprouting up, and it was looking to photosynthesize. It was coming, so we tried digging it out. We think it helped. We're not sure. We spent two hours trying to dig it out."

"We decided to put our tomatoes here because we knew we were going to put them on landscape fabric. And so when we do a new plot, we want to do longer-season crops and try to put them into landscape fabric where the weeds are. To mat the rhizome weeds down so that it's

Credit: Andrew Mefferd

actually doing an occultation process through the season, and then they don't grow through the wintertime," Michael said.

"That's a great idea. One way to break in some no-till beds would be with winter squash. I don't know if that's what you had in mind or something like that where you can put a silage tarp down, cut holes, plant the squash through the holes, and then you set it and forget it. You get a crop off the field and occultation at the same time," I said.

"Yes, that's a good idea. It just depends on your energy levels to put that much energy into burning holes for winter squash, because that's a lot of holes. You can buy them now with holes already in them," said Michael.

"Yes, we got these with the holes already in them," said Shanon.

"We would not have done all those burns. It's too meticulous to do that [with a flame weeder]. And it always seems to burn the plastic," said Michael.

"So we're trialing out the lettuce growing in there. The one thing that is difficult with the landscape fabric is that you have to pull it out. And so there is energy required to pull the landscape fabric back off."

"It's that compared to hours of weeding. What would you rather be doing?" said Shanon.

"It just depends. We have a lettuce crop over there. We weren't used to weeding, and that was our problem. We never weeded. All of a sudden we came here and we looked out and it's getting out of control. I didn't even have a push hoe or anything. And now I've got this stash of weed-eating tools and we're figuring out which we like best. It's funny. You figure out your favorite weed-eating tool in year four," said Michael.

"The Johnson grass has already gone down in a year. I think we can get it down in eight to ten years. I hope that there will be hardly any Johnson grass if we keep on matting it. Bermuda grass, it pops up here and there if you let it get out of hand. Then you can't even get it out. But the biggest one is the bindweed. I don't think we'll ever get bindweed out, honestly. That's just such a vicious little weed and you can't really pull it out and it breaks and it just kind of shatters. But you can see there's bindweed that was in these crops. And by tarping it in the fall here, the bindweed dies off and doesn't grow in the wintertime."

"Little tricks, trying to take the labor out of weeding, are key," said Shanon.

"So the toughest thing that we're finding on these type of weeds with no-till is the borders," said Michael. "The worst border is always the one farthest away from the area you walk in. So the ends of the rows are where our toughest weeds are. And they're creeping into our beds a little bit."

"But you know, the amount of weeding that we actually do compared to other farmers is still way less. I mean, I think I spend about thirty minutes a week weeding. But this is a very small plot—thirty-two 100-foot beds, so around 13,000 square feet. So it's basically walking up and down, just spot-checking. And we're also getting better at weed management, getting them really small."

"One reason why we wanted to have a bigger piece of property is so we can leave plots fallow. We can water them, have the weeds come back, and then flame it down, or cover them. Without disrupting our rotation," Michael said.

Credit: Hillview Farms

A mix of sprinkler and drip irrigation at Hillview Farms' original location.

Permanent Bed Turnover and Maintenance

"The biggest thing is, when those weeds get out of control, how do you go back?" said Michael. "We had the landscape tarp here under growing tomatoes, and we did weed once in between the rows, and then in the summertime we were so busy, we didn't weed much. But look at the amount of weeds in there." A bunch of weeds had grown up over the course of the season through the holes in the landscape fabric where the tomatoes were planted.

"And so the plan is now we have the landscape fabric on the rows. So I'm gonna take out the T posts, I'm going to occultate the area with a silage tarp. In the springtime we'll grow our cucurbits in here, [with the landscape fabric and drip tape left over from the tomatoes the previous season]. Because I'm just going to leave the drip tape in there. If the gophers mess it up I'll just have to leave it, and then run drip tape on top if they chew into it," said Michael.

"So that'll take care of these weeds and then the beds themselves will have been under cover for a year plus. And then we'll see how we want to do the fertility management. We might take soil samples in spring just to see what's in there, and then go from there because we're not going to add compost again. We did a heavy composting and it's already starting the system, so we don't need to jumpstart it again.

"One of the reasons we added compost almost every [time we planted a new crop] at the other farm was because we needed to create tilth to plant into, especially those first couple of years. So if your soil is already softer and doesn't need the extra compost, it doesn't seem like adding compost all the time is necessary. You know there's not just one road map. It depends on the scenario, and you've got to find your way around [your soil].

"So the ideal thing, if we started over again we'd get a tractor that had a wheelbase of at least thirty inches, that that could straddle a bed and drop compost. Right now we basically do a couple beds [at a time], and we'll drive the tractor on there and we'll smash down the sides of the bed and then over time it'll kind of smooth back out again, but it's not ideal.

That's the big thing, keeping the soil covered. Whether it be mulch or crops, disturb the soil the least amount possible.

— MICHAEL WHAMOND

"On the intensified level, the one thing that was really hard for us to keep up with is how fast the system moves. Because you're always transplanting, you're always harvesting. And when you have a four-season model you get burned out. If you don't have that labor to take some of it away. I mean, it's us two and then we have one part-time person that works fifteen hours, and so you want to space it out a little bit more by using the occultation, the dry seedbedding.

"So the initial idea of creating your system in a no-till market garden, is you really want to think about how you're going to lay everything out. Next year we'll have a hundred permanent beds and then the max I think would be about two hundred permanent beds 100 feet long. That would be more than enough. And then once those are set we want to get rid of all the equipment, never use it again. And then we just maintain it."

"We can plug and play," said Shanon.

"I believe that each farm that wants to do this is going to have different challenges. And it is most likely that the first hurdle to overcome will be weed pressure. Figuring out how to combat weeds is essential for a no-till operation," said Michael.

"Let the soil do what it does best. It's trying not to work against anything, I think it is a little bit more just working with nature," said Shanon.

"It's just simplifying things," said Michael. "And it's putting the soil first, that's the key."

Improving the Soil and the Business at the Same Time

"I feel like we're in a hybrid model between Jean-Martin Fortier's model [writer of *The Market Gardener*] and Singing Frogs. We're trying to take the technology that Jean-Martin is using with these seeders and tools [to work more efficiently], and still use Singing Frogs' basic principles of keeping the soil covered. That's the big

Rolling out the compost to prepare beds.

Credit: Hillview Farms

thing, keeping the soil covered. Whether it be mulch or crops, disturb the soil the least amount possible," said Michael.

"We see in the future that we're not going to be able to have human-scale labor. There's just no one around here. So we want to try to do this on a scale where we use technology for labor, and think that's going to be the key, especially with no-till."

Shanon and Michael offered to show me their soil. We walk over to one of the patches that's been in no-till the longest on the farm. The shovel goes right in and there are lots of earthworms in the shovelful that comes up.

"I mean you've got earthworms coming off of the dang shovel," said Michael.

"That's pretty awesome. I want a picture of that," said Shanon (see photo pg. 189).

"On top it's still a heavy clay, but it just falls apart. The crumble and the texture are just perfect. And you can see where our mulching has

A close-up of the mural on Hillview Farms' walk-in cooler.

Credit: Andrew Mefferd

started, and then the soil continues. But the aggregation of it is really good," said Michael.

"Just look at all those earthworms. Employees of the month right there," said Shanon.

"Soil reacts pretty fast. It gets to that stage really fast. It wants to create life. It starts going really fast. We're just trying to manipulate it to get there faster," said Michael.

"When you have permanent beds, you're able to create [hedgerows] that you can bring into the farm a lot closer than when using machinery. That's the next stage in our farming operation, to start making hedgerows and beneficial insect habitats right next to our fields. That was the one thing about Singing Frogs Farm that I realized. I got excited about no-till. I got excited about the profitability aspect. But the emphasis on the hedgerows I think is important," said Michael.

Sprinkler irrigation staked in the pathways.

"On our farm, we've never sprayed one organic pesticide. Looking at our crops, there's stuff eating it. But there's no real damage. A lot of it has to do with timing too. If the greens grow up and they go past their peak, we'll start getting bugs. But really bringing those bigger predators and having homes for them is key.

"It's just the evolution. Once you stop tilling the soil, then it falls into line. You have more time to spend doing other stuff, and thinking about the farm. Maybe it's not just no-till, but I think permanent beds are definitely a key for small growers. They can standardize.

"We haven't talked about money at all in gross sales, but we're willing to talk about that stuff too. We're right in line with the $100,000 an acre. Probably a little bit more," said Michael.

Credit: Andrew Mefferd

"One thing that's interesting is the farm has actually scaled down every year. We started off at an acre and a half. Then we were at an acre and a quarter at the other farm. Now doing these two plots, we're maybe at three-quarters of an acre. A lot of that had to do with time management. But actually we'll be down this year from last year. Last year we did about $115,000 in gross sales. Then this year, I think we'll land about a hundred," said Michael. "But yes, you can produce so much on the small scale."

"Our no-till model has been economically viable for us thus far. Singing Frogs Farm is doing a great job teaching people about no-till farming. We started with the knowledge we gained from going on three farm tours there, and just dove in after that because it really is just that simple. We've changed it a little bit to fit our system," said Shanon.

"When I listen to the Farmer to Farmer podcast and someone says, 'You know, I use 20 tons of compost an acre,' and Chris Blanchard is like, 'Whoa! That's a lot of compost,' I think, 'Holy smokes. I put 40 tons of compost an acre.' If I told him that, he'd say, 'What the heck are you doing out there?'" said Michael.

The sun going down on a field of freshly set lettuce transplants.

Credit: Hillview Farms

"I hope that's something that research projects can give us some clarity on: what compost is actually doing and how much you need. [One of the soil scientists] thinks that just a dusting of compost is enough to, basically, ignite those soil organisms and create that connection. And we're using it as a mulch. I don't know. There are a lot of questions."

"It's an exciting time because the research is happening right now," said Shanon.

"Those Jean-Martin Fortier-style farms, they're doing around 40 tons an acre of compost. It's a tremendous amount of compost. The hardest thing that we're running into is finding quality compost. Because we do the same thing. We start off with about 40 tons an acre [to build permanent beds]. Now we're at year four at the other farm. We used about 15 tons per year. It doesn't need as much anymore," said Michael. "We sourced seven different facilities for compost."

"And we used the best compost we could find," said Shanon.

"At the Auburn farm, I thought that since the earthworms were doing all the work for me over there, I didn't need to broadfork or loosen up the soil. What was happening was that even though the soil is being covered with plants, just the water hitting the soil started compacting it. It was getting really hard to work in there," said Michael. "Now, we do a broadfork at least once a year. My hope is once we get that established and keep on working it, we won't have to broadfork as much, because I don't like broadforking.

"It's a good workout," said Shanon.

"A lot of these things we're trying to change to be efficient with our system. We don't want to use wheelbarrows [to lay compost]. We don't want to do broadforking. It's hard work. And you want to do it as easily as possible," said Michael.

"Yes. Work smart, not hard. That's our motto," said Shanon.

"We use the broadfork and it definitely helps out. We also use more micro-sprinklers [instead of drip irrigation]. That has helped out a lot, because what we're running into was that when you have the bigger cells and you're transplanting, you get a bigger crop to put in, so it covers the bed more and establishes faster, and smothers the weeds," said Michael.

"But what was happening is we were putting the compost on, and some of it was hydrophobic. The water from the drip lines would just bead off. So we would have to water a lot. Even though it was a bigger transplant. And in the summertime, we have multiple days of 100-degree weather. The drip wasn't enough to establish the crop. We'd have to leave it on for 24 hours just for it to soak in. The pathways would turn to rock. What I was thinking was if I do more of the micro-sprinklers, I'm not only growing in my beds, but I'm also softening up my pathways so that roots could grow there," said Michael.

"What we're looking at is not necessarily growing the farm on top, but how do we connect the farm underneath. Having this plot be connected as one city under the ground. We can start geeking out on soil real easily."

Hillview Farm is the change I hope to see in the world: People with no agricultural background getting interested in farming, being able to get started with little land or infrastructure, growing their business, saving farmland that would otherwise be developed, and creating a less centralized, more resilient food system.

ON THE AFTERNOON OF AUGUST 10, 2017, I PULLED up a long, steep driveway to Lovin' Mama Farm. Having spent the morning visiting Four Winds Farm in Gardiner, New York, and the afternoon on the two-hour drive up to Amsterdam, I was looking forward to stretching my legs and visiting this farm. It was going to be a day of contrasts; one of the most established no-till farms I could find followed by one of the newest.

All the way at the top of the hill I found the farmers out in their field, half an acre of newly, neatly laid out intensively planted beds of vegetables and flowers. There was a packing shed, a greenhouse, and a house under construction just across the driveway. We headed into the shed to get out of the afternoon sun.

Corinne Hansch had contacted me after we ran a few stories about no-till in *Growing for Market* magazine. Corinne is from California; her partner Matthew Leon grew up in Manhattan. His family has owned the land they are farming since he was a kid. After farming in Mendocino County, California, close to where Corinne grew up, and losing the lease on the farm there, the couple moved back to New York to restart their farm on Matthew's family land.

"We are using exclusively no-till on our new, intensive vegetable and flower farm here. We ran our own farm for five years in northern California using tillage and lost control of weeds, which in some areas got worse every year. However, we also had a lot of successful trials on many common no-till methods, including sheet mulching, intensive spacings, heavy compost addition, and cover cropping/interplanting. When we made the big move back to family land here in upstate New York, we took a year off to set up infrastructure and process the big transition," said Corinne.

"Paul and Elizabeth Kaiser from Singing Frogs Farm (see interview p. 275), so close to our old farm in Mendocino County, sealed

Corinne Hansch and Matthew Leon
Amsterdam, New York
Mixed vegetables and cut flowers
*Occultation and
applied organic mulches*

the deal during their wonderful keynote presentation at the NOFA-Massachusetts conference in 2017. I had already been doing lots of research and watched some of their talks on YouTube. And after doing their all-day intensive workshop, Matthew was convinced. We are also doing microgreens in the greenhouse, and they are basically floating the farm income while we get our no-till permanent beds up and running."

A diverse array of crops are grown in the no-till system at Lovin' Mama Farm.

Credit: Lovin' Mama Farm

Methodology

"So far, this is the best method we have found," said Corinne. "One, do a soil test at the proper time of year. Then choose soil amendments according to the recommendation on the soil test, while you use a silage tarp to black out [occultate] a mowed area, for about four weeks. Two, mark out beds and broadfork. Three, sprinkle amendments, absolutely essential on new ground to deal with nutrient deficiencies. We are using organic fertilizer for the first time ever, partly because our climate is so mild and the only compost we have access to isn't very high in nitrogen. In such a cool climate there is little heat to kick-start the microbial activity in the soil, so a little fertilizer is crucial for microbial activity. Five, top-dress with four or more inches of compost.

"The compost element has been very challenging, since we aren't making enough of our own yet. In California, we used organic cow manure mixed with rice hull bedding. We paid top dollar to get it trucked in from an hour away, and it completely transformed our farm. Come to find out, here in the middle of dairy country, there is no cow manure to be found. All the dairy farmers use it on their own cornfields, plus it is all liquid. Instead, we are using our local municipality's yard waste compost. It is wood chip based, which is great for mycorrhizal support, but just not doing it for [fertility] our first year on this ground. We are trying lots of different blends, mixing it with peat moss about fifty/fifty, which seems nice so far, especially since our soil is clay loam."

"We are using our small Kubota to [apply compost] down the rows. We're not into wheelbarrowing a half acre plus of compost (we are hoping to grow to three acres eventually). So the tractor straddles the bed and one of us shovels out the compost from the bucket while the other drives down the pathways. We have used some straw in the pathways, but had some weed germination from seed in the straw, so we're always on the hunt for good weed-free straw," said Corinne.

Credit: Lovin' Mama Farm

Dense plantings shade out weeds.

How It's Working Out

"So, how is it working out? There are goods and bads. Goods being that we could get in the field as soon as the thaw came in mid-April. It's been such a wet spring that we would probably have been twiddling our thumbs waiting for the soil to dry out enough to till. We've found that when we add proper fertilizer and necessary amendments (after a soil test) and a thick enough layer of compost, there are no weeds and great growth," said Corinne.

"We've already flipped the first few beds, and love having transplants at the ready to go right in. Direct-seeded lettuces, beets, cilantro, dill, carrots, radish have all germinated okay. Also, the crops grown on top of silage tarp occultation areas seem to be doing much better, and the weed pressure is less.

"When we didn't amend with fertilizer, it was not so good. When we didn't add a thick enough layer of compost (so some native soil showing), there were lots of weeds. We are starting super small our first year so we can stay on top of the weeds and keep our fingers crossed for weed-free beds from here on out. Using weed-free compost and pathway mulch are important lessons, too."

Why No-Till?

"We did five years in California, with tillage-based farming. Tillage is amazing from a work perspective, because all of a sudden you've got a third of an acre prepped, and you go in and plant it all. You have to pay the consequences for that instant soil prep, though. The weeds come in like crazy and the soil loses tilth and gains compaction. And especially over time, we noticed the weed pressure getting worse and worse and worse, and we were eventually losing whole crops to pigweed. One of our main drives for doing no-till is weed suppression. The economic drive is also huge. Seeing what other no-till farmers are doing economically is pretty exciting," said Corinne.

We walked over to the first-year market garden, which consisted of forty-five 100-foot beds.

"I'm okay with having just this many beds. It's been a little hard to scale back after growing four acres at our old farm. Starting from scratch here, we really want to pace ourselves and do it right from the start," said Corinne.

"It's a learning experience, too. We took that really seriously, because we care about the Earth so much. We were tilling because that's what we could do with four acres. The two of us farming, we felt we needed to till to maintain that much space," said Matthew.

"We had a newborn baby when we started our farm in California, and two other kids. Our third baby was born our first year into farming full-time," said Corinne.

"But now we switched to no-till for many reasons. Mainly, we want to grow food in ways that regenerate a healthy planet. I want to do this for the Earth, to help heal the Earth, and it works so well! We've evolved our growing methods to address the degradation of the planet due to poor agricultural practices," said Matthew.

We've evolved our growing methods to address the degradation of the planet due to poor agricultural practices.

— Matthew Leon

"It's definitely a big experiment this year," said Corinne. "For years, we've been trying to do research about no-till. But we were finding it very hard to find any information about it, and how to do it. We were in Mendocino County, which is just one county north of Sonoma, where Singing Frogs Farm is. I had been trying so hard to get to see them, but never was able to make it to their farm. It's funny that we didn't get to see them until we moved east. We saw them at the NOFA conference in Massachusetts last winter.

"And the Armours [of Four Winds Farm, see interview p. 145], they were one of the only farms who had any information online. They had this little video on YouTube that we watched over and over again. So, we're actually copying their method. We drive over the beds with our tractor. We're in our late thirties, and not really into wheelbarrowing compost around that much anymore. Been there, done that. So we're using the tractor with the bucket. We have one person unloading

On the summer day I visited Lovin' Mama Farm.

Credit: Andrew Mefferd

from the bucket, and another person driving. Or he can do it solo, too. Just park the tractor, unload, drive it up a bit more, unload. Way less backbreaking. We're happily using the tractor for that."

"I also know there are some drop spreaders that could increase the efficiency for mulching with bulky material like compost," I said (see Resources section).

"Kind of like a manure spreader, but it just drops it," said Matthew.

"That would be nice because you could drive over and it would just drop it right in the row. We've been dreaming of designing a spreader like that,'" said Corinne.

"I have dreams, 'Oh, maybe someday I'll have a roller-crimper, and I'll be able to do a corn patch.' Or a big patch of whatever," said Matthew.

The Origin of the System

"We started using this method years ago, before we were even farming. We were just gardening. It came naturally to us. After I graduated from college in 2001, we moved back to my parents' place, which has a large garden. I grew up doing farmers market with them. It's super small scale in a tiny little town in northern California," said Corinne.

"And we grew a farmers market garden," said Matthew.

"At the end of the season, we bought all this horse manure, and we just top-dressed a thick layer onto the beds. Then we moved to our next adventure. The next year, I remember my mom saying, 'This is the best garden we've ever had. There are no weeds. We don't need to till.' A few years later, we were working on another guy's farm in northern California, same thing. We just top-dressed and planted right into it. They said, 'What are you doing? You're not tilling? This is so weird.' We said, 'Trust us. It's going to work,'" said Corinne.

"It's really hard to listen to those instincts at times, to follow your gut feeling when no one else is doing it, you haven't seen proof of it happening. So I'm really grateful to Elizabeth and Paul Kaiser of Singing Frogs Farm for getting word out there and creating a movement. They're not the only ones. There are lots of others doing it," said Corinne.

"Where did you get the idea for the methods that you're using now?" I asked.

"I remember being up in Olympia at the Evergreen State College. I worked on the organic farm there. One year we planted garlic, and I top-dressed it. I remember when I pulled the garlic out, the roots had gone up into the top-dressing. They were these massive roots. That made an impression, and top-dressing just seemed like a big part of that," said Matthew.

"For me it was watching some of Singing Frogs' YouTube videos. They have all these ideas about how no-till relates to climate change. Then they get into the nitty-gritty of how they do it. That's the kind of information I am just voracious for. Learning from other farmers and seeing how they're doing things. Gleaning the things that make sense and work for us," said Corinne. "From the Armours, we got the idea of just driving the tractor over the beds to put down compost." said Corinne.

Credit: Andrew Mefferd

Smoothing out the compost on top after flipping a bed from a previous crop.

"When we were in California, I took a workshop up at Bountiful Gardens, through Ecology Action, with John Jeavons and that whole community up there. We were walking around their garden, and I noticed that what they do to clear their beds is they just pull out the crop, put it in the compost pile, they hula hoe it, and then they plant again. They were like, 'Why do you want to till when you can just hula hoe it?'" said Matthew.

"And when you're hula hoeing it, you're getting out the last weeds and cutting up that stuff. It disturbs the soil a little bit. Then you can plant into it. I think they would amend [after that]. That was a little green flag for me."

"Also in California at our old farm we did a little bit of no-till with some of our strawberries and raspberries. We stopped tilling and just top-dressed. It seems counterintuitive. I think that's why a lot of people are so skeptical. They think, 'Oh, the soil is going to be hard and impossible without a rototiller or harrow.' But it's actually softer and fluffier. You have to really see it to believe it. When we were hand weeding in our no-till strawberry patch, the weeds were just coming right out. It was so different from the onion patch that had been tilled, and it was rock hard, impossible to weed, even though we'd amended it with compost," said Corinne.

Just beyond the current field, a new section of field is prepared by occultation with a silage tarp.

Credit: Andrew Mefferd

"It is a leap of faith going to no-till. I have talked to so many farmers that say, 'If I can't bring my tractor in and just till it, then I'm screwed.' In response, well, yes, you will not be able to clear weeds, blend in compost, and be left with a soft top layer of soil with the tractor. That's what I think a lot of farmers really want to see and why they have trouble adapting to no-till. They want to see what they put down on the soil worked in. And they want to have a really soft soil that they can just push transplants into," said Matthew.

"Yeah, there've definitely been times this year when we've been tempted to till," said Corinne.

"The tillage and other deep soil tractor work like that, it's superhuman and should be used sparingly, for instance when breaking ground the first time only," said Matthew. "Sure, machines can do deep work, and we need to be careful about how deep and when we do this work."

"You can do it with roller-crimpers on the large scale, and that can be a pretty great way to do it, that doesn't till and break up the soil and then have it wash away. There are ways to do it on the large scale. However, if you want to pump out a lot of food in a small space, it seems like what we're doing is what a lot of farmers have turned to, about a thirty-inch-wide bed that has a bunch of plants in it, that shade out the competition of weeds."

"We also have the microgreens business, that is floating us right now. So we didn't feel this pressure to pump out a ton of food right away. We felt like we can experiment with some no-till techniques. We just kind of went for it. It's been going really well, and we actually are pumping out a lot of food!"

"My father owns this land. As a kid, we came up here a lot of weekends in the summer, and for weeks at a time. So I grew up coming up here a lot. I've got a deep connection with the place. It's wonderful to see my kids running around here, and experiencing the same kinds of things that I experienced when I was a kid. Now to be making a business and a new home here, it's pretty awesome. We're really excited to sink our roots down and not have to uproot them again," said Matthew.

"This is our first year here, so there's still so much to learn. We've already, in just flipping one round, we learned so much. I know that so

Tomatoes planted on landscape cloth.

Credit: Lovin' Mama Farm

much more learning is going to be coming. Like spacing, for instance. You can really pack things in so much more in an intensive no-till system. I used to always do peppers in a foot-and-a-half or two-foot spacing. I experimented a few places with one-foot spacing and realized I can get great production with my peppers, okra, tomatoes, even broccolini and cabbage too," said Corinne.

"It seems like if you have enough fertility, you can probably pack things in more densely than the conventional wisdom would allow," I said.

"Yeah, and then they shade out the weeds, and they create this whole canopy of support for each other," said Corinne.

Refining the System

"Something that we're still working on is the direct seeding in this system. I think part of the challenge is that we don't have an irrigation system yet. It's been a great year for not having an irrigation system, with all this spring and summer rain," said Corinne. "We're still adjusting to rain in the summer after five years of farming in dry California. So with a new growing climate and a new growing system, we haven't quite figured out the seedbed thing."

"We have a Johnny's six-row precision seeder. If your bed isn't the perfect consistency, as far as tilth goes, then it binds up. If we just put down compost and rake it, and you try to roll that thing over it, it just binds up with the soft compost. Suddenly you're not seeding well, and it doesn't really work. So, we're trying to figure out how to make that happen," said Matthew.

"Meanwhile, we just switched over to the Salanova lettuce [which is for salad mix but transplanted], and we're transplanting everything. We even transplanted beets. First time we've ever done that," said Corinne.

"We do a lot of transplanting. The thing is that you can get the plants exactly where you want them," said Matthew.

"You get the density that you want. We do the Salanova every six inches down the bed, four across. So you've got eight plants to cut out every foot versus a million with the precision seeder. Then when you're cutting the crop out it's a lot easier to flip the beds that way," said Corinne.

"We're following the Singing Frogs Farm method. When we flip our beds, we go in with knives. We cut the previous crop out right below the soil, each plant. We put the past crop in the

Compost is applied to the top and straw is applied to the pathways when remaking beds for a new crop.

Credit: Andrew Mefferd

compost pile, weed the bed real quick if necessary, amend and top dress with compost/peat, and replant." said Matthew.

"I think perennial weeds are another fear that people have around no-till. Definitely, as I was reaching out this winter to different university researchers, saying, 'What do you know? Where can I find more info about no-till?' Most of them responded with concerns about invasive, perennial grass weeds. Which so far, we haven't seen any," said Corinne.

"I think one of the big things about the no-till system for us is that, we're used to tilling up a sixth of an acre and seeding one of those every month. Then sometimes it's May or June and you've got to plant a half an acre, because all the tomatoes are going in. With no-till we realized that we really need to keep down the amount of space that we're trying to cultivate at once. At least at the start here. Because everything is by hand at this point," said Matthew.

"You can see where we laid down this hay, there was some seed in the hay, and suddenly there are weeds growing in the pathways out of the hay. Then, we have to weed that. And our beds are pretty weed free, but still we're constantly trying to go out there and make more beds. It gets overwhelming. Suddenly you're like, 'Wow. I only have so many hours in my day.'

"And as much as we want to expand more, I feel like we shouldn't, because the worst thing that could happen would be that you let weeds grow up in your patch, and they go to seed. Then you're laying down this compost, and then they're going to seed on it."

"We have forty-five rows out there, 100-foot rows, so about a half acre, and we've got our propagation greenhouse. We started marketing at the very end of February with our microgreens. It's looking like if we can continue the average we've had over the year, we'll be grossing around $70,000 this year. And it's just the two of us," said Corinne.

An update after the end of the season showed Lovin' Mama did even better than projected over the summer, grossing $90,000 off of their half acre of no-tilled flowers, vegetables, and microgreens.

"That's a great first year," I said.

"Yeah, we're doing back flips. That's amazing. It took us four years to get to that point in California," said Matthew.

"But we were in a very rural area. There are just so many more eaters here," said Corinne.

"We're right next to Albany and Troy," said Matthew.

"The marketing potential here is just light-years beyond what it was in Mendocino, thanks to the population," said Corinne.

Starting with Experience and a Plan

We left the barn to look at the field. Neatly laid out 100-foot beds had straw-mulched pathways. A border of straw bale flakes surrounded the field, as a form of long-term occultation to try and keep grass from encroaching into the field.

"The other thing that I think about as we're standing here, is that we've found ourselves in a unique position. Having five years of farming experience, and then landing on this new site, we took the time to create an intentional plan around field layout and infrastructure development. We spent the first year (2016) starting a small microgreens business," said Corinne.

"We plowed the field," said Matthew.

"Yeah, we did plow to break ground. There were some really thick grapevines, and raspberries," said Corinne.

"Goldenrod everywhere," said Matthew.

"So we mowed it, and then we got a plow, and we plowed the whole thing. Then we planted rye," said Corinne, gesturing to the rest of the field outside of their straw-bale border.

"We put a lot of intention and thought behind the layout. We've got four zones in a three-acre field, all with 100-foot beds. This is going to be our 100-foot greenhouse, which fits perfectly in zone two. It's also right next to the packing shed so it was more cost efficient to get power, gas, and water out, and it's easy to run the harvest in," said Corinne.

"We're still in the stage of farm development here. So, just taking your time with the things that you can take your time with. What can we not take our time with? We need to make money. We need to get the greenhouse up. We need to get the packing shed up, so we can wash things. And we need to start farming the field," said Matthew. "The infrastructure comes little by little as we have time and money to invest."

"An intensive no-till system helps make all of that possible, because you're getting so much production with so little space. You can slowly build up your farm to a point where it's producing what you want it to be producing, and to get all those systems in place. It takes time. We know that from our experience in California," said Corinne. "We definitely knew we wanted to grow smarter, not bigger."

"And we've got plenty of space in this field, but it'll be there. We don't have to be expanding so quickly that suddenly it's out of control. And saying, 'Oh, we don't have a crew to weed it all, so we're in trouble,'" said Matthew.

"So the pathways have been a really interesting evolution," said Corinne.

"Yeah, right here, you see where we put down hay," said Matthew, gesturing at the pathways between beds. "We had to weed it, because it was growing grass. I got pretty upset about that. It's one of those things, I thought, 'I put it down to keep the weeds back, and it's growing weeds.' Germinating oats or whatever it is in the hay. So now, we're spending the money on straw, which is harvested by a combine. It doesn't have many weed seeds in it. We'll see how that goes," said Matthew.

"I'm hoping that eventually the straw will break down and suppress weeds, and we'll get to the point where we don't have to keep laying it down," said Corinne.

"So you have weed seeds that blow in, but hopefully you take care of those. If you can stay on top of that, and keep any more plants from going to seed in the beds then your weed pressure should go down over time," I said.

This interplanting of brassicas and lettuce shows how very few weeds have sprouted since the crop was planted. The lettuce will be harvested before the brassicas shade it too much.

Credit: Lovin' Mama Farm

"Exactly. These onions were our first round. This is an example of some beds that we did not put down enough compost on. It was our very first experiment, and you can see the weed pressure there. We even weeded them once by hand. But then these beds over here have been flipped at least once," said Corinne, pointing to several weed-free rows.

"We added more compost. And we realized that we needed to add amendments as well. A lot of the things we planted in the first round were not growing very well at all, and it looked like a nutrient issue. So we said, 'Okay. Time to go get some bagged fertilizer from North Country Organics.' After that we cut out that last crop, amended with the fertilizer, and top-dressed with a thick layer of the compost/peat mix. The next round of crops grew well," said Matthew.

We walk by flowers, Brussels sprouts, okra, and a wide mix of other vegetables.

"It's probably a good idea to just have one crop in a 100-foot bed. That way when it's all done, you get to cut it all and pull it out and flip the bed. And you don't have this," said Matthew, gesturing to a partial row of egg-plant.

"We have a few beds over here with a little bit of eggplant, because we were just desperate in the spring, to squeeze everything in. I had started way more things than we could ever fit. It's hard to transition from four acres of production to a half acre!" said Corinne.

"It was amazing, because the spring was so wet, even into July the field was probably too wet for tillage. We don't have any drainage tiles or anything like that in here. So if we had been waiting to till to get in, I don't think we would've gotten planted. That's why the pathways got away from us. We were plan-

The silage tarp shortly after it was applied.

Credit: Lovin' Mama Farm

ning on hoeing, but it was so wet. You would hoe and they would just come back," said Corinne.

"And getting those early cucumbers and zucchinis and tomatoes is important. That's when people are crazed for them. Then you hit the August glut. Not sure yet if that happens here, but definitely in California, everyone in Mendocino has a backyard garden with cucumbers and zucchinis and tomatoes. And then they don't want to buy any," said Corinne.

Occultation and Mulching with Compost

We walk over to a tarped patch of ground.

"This whole section was tarped with a 50' × 100' silage tarp. We left it on for four weeks," Corinne said, gesturing to a dozen beds directly adjacent to the tarp.

"And weeds aren't growing there," said Matthew.

"The earthworms go crazy under the tarp," said Corinne.

"You had it tarped for four weeks during what part of the year?" I asked.

"Mid-May to mid-June," said Corinne.

"Are you happy with the amount of weed suppression? Can you even tell? So you pulled the tarp off and immediately built beds here? It looks like there are not a lot of weeds coming up through," I said.

"Yeah, I think honestly, the amount of weeds that came up after we pulled the tarp off, it was mainly just Virginia creeper and some other really deep-rooted weeds that had some life down there still," said Matthew.

"Let me just make sure I'm getting this straight. Your method is to do occultation. And then build a bed with your various compost materials. Is there a depth that you're shooting for? Are you just eyeballing a few inches?" I asked.

"I'd say four to six inches. Maybe less. Probably more like four," said Corinne.

"We make a pile down the middle, and then we rake it out. The pile is probably six inches deep," said Matthew.

"We did broadfork, too. Which, I don't know that we'll continue doing. Maybe once a year, every now and then on beds that seem like they need it," said Corinne.

"Adding all this straw, is this something you're doing in the first few years to try to kill the stragglers? This isn't a long-term strategy, is it?" I asked.

"We don't know. We're experimenting. My hope is that it's not long-term. If we can really stay on top of keeping all the access rows mowed, and all the edges mowed. We also have a really cool sheet mulch project that we did down there with our lilacs and peonies," said Corinne.

"And we're starting another patch right here. We're going to lay down the cardboard and put wood chips on top of it," said Matthew.

"And just let it sit. It's going to be a lot of shrubby perennials, I think," said Corinne.

"When we do plant that area, we'll just decide where we want our plants, push aside the wood chips, cut a hole in the cardboard if it's still present, and just amend that one spot and put in a plant right there," said Matthew.

The field at Lovin' Mama Farm with the new house behind it.

Credit: Andrew Mefferd

"The monarchs are just everywhere, there are so many butterflies right now," said Corinne.

"Yeah. That's the other big reason to have flowers in the fields, all the beneficials," said Matthew.

"That's another great thing about no-till that the Kaisers are so big on is creating habitat for those burrowing beneficials. One of the other workshops I went to at the NOFA-Mass [2017] conference was on native pollinators. I'm really interested in that. There are a lot of native pollinators that are ground burrowing. When you till, you're destroying their habitat. We want all pollinators," said Corinne.

We stand back and look over the tidy rows of flowers and vegetables.

"Our little tiny farm. It's pretty awesome. We're going to have our house right there," said Corinne, pointing to the house-in-progress right by the field.

"Yeah, that'll be nice. I like your commute," I said.

"Yeah. A few steps. We're really blessed," said Corinne. "We're excited for all the potential things we can do to revitalize our little community here. Our little economic input. There are just so many great things about no-till. It's environmental, economic. We want to invite college groups out. We definitely have a drive to spread the word. And grow good food for our community," said Corinne.

"That's one of the things that excites me about this. It lowers the bar a little bit, as far as people not necessarily having to buy a tractor. You can start smaller," I said.

"We're definitely holding on to our implements for now, until we figure everything out," said Corinne.

"That might be a message there for people: Try it, and hang onto your tractor. Hang onto all your equipment. Try it out, and you can always go back to tilling if necessary," I said.

"Yeah. No need to abandon your equipment," said Corinne.

Suggestions for Getting Started

"You're in the process of starting this. Any tips? Is there anything else that you would want to say to someone who's thinking about giving it a shot?" I asked.

"I feel like the collective wisdom of people who do no-till is that once you get it set up, it's great, because there's not much weeding, and you get high production. You can make a good amount of money on a small amount of space. We're finding that true in our first year, which is when the work is the most. The weed pressure is still there. We decided to take that jump, that leap of faith, and we're already finding the benefits of a no-till system. We're weeding less than we ever have. The weeds aren't overtaking us. We feel like we're keeping it maintained," said Matthew.

"I was going to echo what you said about just trying it. Because we have the microgreens, we knew we were making a certain income from them, so that made it a lot easier to try it. If you know your tillage system will give you this much money, maybe maintain that so it's floating you while you're experimenting with no-till. Don't be afraid to experiment. For me, that's the joy of farming. Always trying something new and experimenting, and building off that knowledge every year. But if you can have that economic security with a system you already have in place, that's already paying the bills, it gives you freedom to experiment," said Corinne.

"One thing I've thought a lot about too out here is that I can build a bed, cover it with weed cloth, tarp, or sheet mulch, and just let it sit for a while. Then when I'm ready to plant it, pull it off and plant it, or plant straight into the sheet mulch for perennials. There are so many ideas out there. There's so much room for variation in this system. I think that the more you keep the weed pressure back, using mulch of some sort and occultation, that you could really come into a system that's highly productive and not as much work. The main things you're doing in this system are bed prep, planting, and harvesting," said Matthew.

"Isn't that a lean farming principle, too? Spending more energy doing the things that make you money. Weeding doesn't really bring you any income, but harvesting does," said Corinne.

And with that, I left them to their farm chores for the afternoon, and took a few more pictures on the way out, feeling very inspired about the success that Lovin' Mama Farm is having.

WHEN I WAS VISITING WITH ELIZABETH AND PAUL Kaiser of Singing Frogs Farm in California (see their interview p. 275), they suggested other no-till farms that I could talk to. They mentioned a couple that worked on their farm and had taken their methods back to Australia. Unfortunately, a trip to Australia was not in the budget. However, as the writing progressed, I found myself more and more curious about how no-till practices from California were translating to Australia.

MOSSY WILLOW FARM

Mikey Densham and Keren Tsaushu
Main Ridge, Victoria, Australia
Mixed vegetables
Occultation and compost mulch

That's how I found myself Skyping at 3:30 one morning with Mikey and Keren of Mossy Willow Farm. With a fifteen-hour time difference, the only waking times we had that overlapped were the very beginnings and ends of days. Good thing I'm an early riser. I groggily sipped my coffee and marveled about how I could be looking face-to-face and talking in real time with two people on the other side of the world.

To be more exact, Mikey and Keren's farm is on the large Mornington Peninsula, a little over an hour's drive south of Melbourne. I was excited to hear how starting a farm with no-till methods was going for them.

Interest In and Learning About No-Till

"So how did you get from Australia to working on Singing Frogs Farm in California?" I asked.

"Through an article we saw online called 'The Drought Fighters'," said Keren. "Mikey and I were looking for places to get a good internship and mentorship, and suddenly this place popped up with a similar climate to ours, a Mediterranean climate similar to Israel and parts of Australia.

"And for me personally, understanding the benefits of no-till came from talking to Paul and Elizabeth. I had no idea of the connection between tillage and the release of greenhouse gases. That was huge for me."

"It was Singing Frogs that kind of threw us into the deep end of no-till. But I remember a mate of mine in my early farming days who gave me the Fukuoka book. So I read *One-Straw Revolution*. That was where our roots of farming came from. Fukuoka was talking about trees and orchards, and I was thinking, 'How can I farm in a natural way with vegetables?' I read the book back to front a few times. He talks about some form of tillage that he does for growing vegetables. I was thinking, how do I do this? So I asked around, and most of my farming mentors looked at me in a weird way and said, 'I don't know if you can really farm without tillage,'" said Mikey.

"We both started farming in Israel, which is a bit of a tricky place to do farming in a nonconventional way because there's very little information about anything that is not commercial, large scale, and usually for export. So the initial seed for no-till farming was in us, but the practical knowledge and facts came from Paul and Elizabeth, and reading a lot after because we got so passionate and excited about what they were doing," said Keren.

"So you guys went to California specifically to work with them?" I asked.

"Yes," they replied in unison.

"And you worked on their farm for a while?"

"For three and a half months, full on. We lived with them on the property, so we were living and breathing it for three months," said Mikey.

"We would have stayed for longer, but I didn't have the visa to do so," said Keren.

"And now you've started your own farm in Australia using their methods?" I asked.

"We're on the start of our journey. We have been on our own property for a year, last summer we were just getting started. It was a tiny CSA, and this season was really the first season of full production," said Keren.

"I'm really interested to talk to people at your stage, because one of the reasons I'm so excited about no-till is because it helps people to be efficient on a small piece of land. I think that's one of the great potentials of no-till is to get more people to try farming if they don't have to get so much land and don't have to get so much equipment," I said. "One of my questions is about how the getting started process is going. So you worked with Singing Frogs for a few months, and then did you more or less absorb their method and say, 'Let's go do it?'"

"We absorbed their method. As Keren said before, we were young farmers. I had previously worked at Brooklyn Grange farm in New York City. It's an incredible rooftop farm. So I had a market gardening foundation, and then Keren and I also worked together on a farm before we went to Singing Frogs. And we ran a small CSA," said Mikey.

"We had just started hearing of Neversink Farm (see interview p. 247) doing no-till, from *Growing for Market*. We thought, 'Whoa. This guy's calling it no-till but is doing it a completely different way.' So we thought, 'All right, there is some nuance to this no-till thing.' But we felt comfortable enough, at that stage, to go ahead and start our own venture.'"

Credit: Mossy Willow Farm

An aerial view of Mossy Willow Farm.

Getting Started

"So you got your piece of land, and how did you start with the no-till system?" I asked.

"When we first got the piece of land, there were about ten oddly shaped home garden beds that were already there. So I had something to kind of play around with while we began thinking about how to create the proper farm plots. The first striking thing was that we were on quite a heavy clay, and I thought, 'How do I do this?' You know Singing Frogs are on much more of a loamy soil, and I remember working in their clay patch at the bottom of the farm, and that was hard work. That was a different ballgame," said Mikey.

"So I was thinking, 'Really, how are we going to manage doing this no-till method on a completely different soil profile than what Singing Frogs are doing?' We had a rototiller on the farm that a mate of mine had just bought, which I got him to sell. I said to him, 'I don't want it on the property because it's going to tempt all of us to use it. He's said, 'Are you sure? We just bought it.' I said, 'Man get it outta here otherwise it will surely be used by someone.' In the end, however, I came to the conclusion that it was going to be rough going trying to work the soil without some sort of tillage. The soil was hard and compacted and covered in turf."

"Also, we didn't have time to put tarps down for a year and wait for the grass to die. So we decided to do an initial till to get the ground opened up," said Keren.

"We called a guy up the road who has a tractor service, marked out an initial plot, and he tilled it up. I remember him saying, 'How many passes should I do?' And I said, 'Let's stick it out with maybe just doing the single pass, and come back once more.' And he said, 'But it's not going to be a fine dust.' I said, 'Let's relax mate,

Beans coming up in the field.

Credit: Mossy Willow Farm

a single pass will do.' Because he was an old-time farmer he thought I was crazy. He was laughing his head off," said Mikey.

"So that was a single pass with a moldboard plow? Or a disc, or what?" I asked.

"Just a big rototiller. That was the only thing he had. No bed former on the back, he just went in there and tilled up the area that we marked out," said Mikey.

"What did you do next?"

"For the first two plots we were a bit stressed on time because we got the opportunity on this land late in the season. So we went in straightaway. We spent lots of energy raking the big chunks of grass out, which was a workout, and then formed the beds. The next two plots we did a bit a later, they were tarped for about a month and then formed. With more time on our hands we could invest in a longer and more efficient bed-forming method than in the first plot," said Keren.

"Yes, initially we were forming the beds, and including a lot of compost. Everyone advised us to invest in compost and take the opportunity that we tilled in order to deeply incorporate the compost before we started layering it on top, so we did that," said Mikey.

"So you tilled the compost in, or just put it on top of where he rototilled?" I asked.

"We put it on top and forked it in manually," said Mikey.

"Did you have raised beds at that point?" I asked.

"Yes, at first. And the original decision to do so reflected our context of having pretty wet winters. But after a while we understood they were a bit too high for us. In the next few plots we stopped digging the path and just put the compost on top of the beds, and then they weren't raised very much at all," said Keren.

"Initially we were digging pathways and putting that soil on the bed. And that was much more of a reflection on how we'd farmed before and also at Singing Frogs. Currently however our bed forming practice is with absolutely no raising apart from just compost application, and the broadforking, which helps to shape the bed. So no digging out paths, and not really raising the bed up any more than we need to," said Mikey.

"I'm just trying to picture how big your beds were. How much compost did you put on: four inches, six inches?" I asked.

"Let's say a wheelbarrow is 60 liters (15 gallons). We were doing up to ten wheelbarrows for a 100-foot bed. So that's, let's say, around 600 liters (150 gallons, or 30 five-gallon buckets) of compost for a 100-foot bed. Do you guys work by liters like that?" said Mikey.

"No. We use gallons, but everybody else in the world uses liters, so I'm used to doing the conversion," I said.

"When you put on all that compost, did it settle down over time? Now are you pretty much growing on the flat?" I asked.

"No, it's still mounded a little bit. Also, because we use wheel hoeing in the paths, the hoeing naturally gives delineation to the beds. I would say maybe like five centimeters (two inches). It's not completely flat, you can definitely see the beds but shoulders are kept to a minimum height. It's just not mounded high enough to create wasted shoulder space," said Keren.

"The old plots that were raised, we transitioned out of that. The shoulders were drying out. And in the initial year we realized quickly that we didn't have the time to manage the weed pressure colonizing the exposed edges of the bed. We understood that by lowering the beds down and reducing the surface area of the shoulders there would be fewer weeds, and so that's why we moved into the new bed shape. When we began the farm, the initial reason for raising the beds was because we were on slope in a rainy area, so we were thinking raised beds would be good for drainage and would prevent erosion. But considering we're in clay soil, which holds together well, and we're on a slant, drainage and erosion haven't been a problem, so sinking our beds down to near flat has been a great decision," said Mikey.

Dealing with Weeds

"On the topic of weeds, and I will make note again that we haven't been farming this property for long, but I can see a dramatic change even in our oldest plot, versus the new plot that we tilled about a month ago," said Keren.

"Now that we have the time, we tarp the area beforehand to kill the grass off, and then do the initial till. This reduces the chances of grass taking root again after the tillage or the need to rake out large chunks of grass afterwards," said Mikey.

"The amount of weed pressure that I have to deal with in the new plot is something that we haven't gotten used to managing. The older beds, which are now at least a year old and have settled into our no-till management, have far less weed pressure. You know there are always weeds. But they're something you can hoe once or twice in a crop cycle and you're pretty much good to go. But in the new plots that we just tilled, it's insane. There are continuous waves of new weeds germinating until they eventually settle under our management. It's a good reminder of why we do what we do. Because if we had that weed pressure continuously on the whole farm we would never get anything done," said Keren.

Harvesting spinach.

"It's management of weeds, as opposed to a war on weeds," said Mikey.

"It sounds like you're going to stick with that way of breaking in beds? You said that even the bed that you tarped you got rototilled?" I asked.

"It was just rock hard. We're on heavy clay and the property was pasture for years, and with just the both of us, there's no way that we could have, just time-wise, opened it up,' said Keren.

"That insight is valuable, because I didn't realize how compacted that was," I said.

"We knew our farm is going to be a certain size, and that once we open a plot, it's just one time of doing that. Then the decision was easier to make, to till it, put the effort in, create the beds manually, add lots of compost, which is a lot of labor, and then know that the plot is pretty much done; no need to re-till or reform each plot every season, the hard work is done at the beginning" said Keren.

"Yeah, we say to people that we're founding a plot. We're establishing it, we're investing in a long-term process of soil creation. So like Keren is saying, when you open a plot, you're going to deal with that heavy war on weeds and the bed-forming work in the first six months. But it's nice knowing that I'm founding this plot in a no-till way and our primary goal is to nurture and create soil fertility. In the long term it eases up and the system finds its balance and becomes easier and more productive. That's a justification that can help you get over making the decision to make the initial till and the ensuing battle against the weeds," said Mikey.

Field Management

"So tell me, how do you make the transition from the end of one crop to a new crop?" I asked.

"We take a knife, cut [everything] three centimeters [one inch] under the surface, put on the compost and then we broadfork everything. We used to put just chicken manure down, but now we use an organic fertilizer, which is a mix of manure, seaweed and more. We then put a thin layer of compost on top, depending on the crop that comes after. So if it's a heavy feeder, compost will go on, and if it's a light feeder then we leave it with just the amendments. Then we rake and shape the beds. Over the past year, however, as our context has evolved, we have begun to slightly change the final steps," said Keren.

"Context is everything. Here at Mossy Willow we rely much more on direct-seeded crops, mesclun mixes, radishes, mustard mixes, carrots, etc, than where we learned no-till farming. We realized our context was very different. For starters, we are a smaller operation with less nursery and general field space. We struggled initially to access quality compost and high-grade potting mix so had difficulty basing our system on large transplants. And finally we began using specific tools, for example the paper pot transplanter, to help cut labor time because our operation is based upon less people. So while we had a toolbelt of incredible skills from our mentors, we needed to adapt to our conditions.

"The biggest challenge came from our experience with our clay soil, which after a long crop and the use of drip irrigation became quite crusted and hard. The hardened surface made it more difficult to transplant. Transplanting takes way longer when the soil is harder. So with our big crops (broccoli, cauliflower, etc), we still cut the crop out, broadfork first, compost, then plant. However, since we began using the Jang seeder and the paper pot transplanter, we had a bit of a problem with our turn-over process and bed preparation. The tools didn't work nearly as well because the soil was too rough."

"Initially, we would apply compost after every crop. That was something Paul gave us advice on. He said, 'When working no-till on a clay soil, farm above the clay.' So we really took that to heart and it was incredible advice that I would pass on. In my mind I'm always working towards growing our organic matter in the upper layers and letting biology work the lower layers of clay into something more friable," said Mikey.

Produce sales at the farm.

"I think the difficulty with that long-term compost application is the cost. You have a growing input cost when you're applying compost every single time. Not to mention the physical time it takes to apply it."

"And also we didn't want to have any [nutrient] leaching or anything like that," said Keren.

"So we were thinking, how do we need to alter the system slightly, to use these tools and continue to obtain high germination and successful crops?" said Mikey.

"Which led us to getting the Tilther," said Keren. "From the first moment I couldn't do it. I felt like it's cheating. I didn't want to do any kind of tillage and I didn't know what the biological and physical consequences were going to be on the soil. On the other hand, our other

Credit: Mossy Willow Farm

option would be to break it up manually with a rake, which, I guess if you think about it mechanically, is a pretty similar process, only with the rake you're doing it by hand, So at the moment we are selectively using the Tilther for beds that will be paper potted or direct seeded."

"I'm surprised that after putting that much compost on top it hardens up that much," I said.

"Also, we're using drip. I was farming in Israel in the Arava desert. I was living there for four years, after the army my first job was farming in a massive pepper farm. We were exporting to Russia and Europe. So my introduction to farming came from the reality of desert farming. The whole thought of water conservation and drip, instead of overhead spinklers, was so strong that when we started here we decided to use drip. The only original area with sprinklers was the salad plot," said Keren.

"We have a designated area where we grow our greens, turnips, baby kale, basically the direct-seeded crops," said Mikey.

"And this has been a really interesting thing for us to witness; how the soil profile in the two different areas of our farm, the beds under drip and the beds under overhead, have evolved. We have been challenged with seeing how under the summer sun, the soil cakes and large clay clods form. As a result we really began to ask ourselves, how can we manage the transition from crop to crop? How can we efficiently transplant into hardened soil? How do we effectively use the Jang and the paper pot transplanter in this tough soil?"

"The salad plot, with the overhead, has been very interesting to witness under no-till. We're seeing there a very different soil tilth and profile evolving. In the overhead area we're seeing far less caking, far fewer clumps, and an increasingly loamy soil developing. In this plot, every single time we take out a crop, we fertilize, quickly tilth if necessary, and then seed again in a matter of hours or in the worst case a day or two.

"So while in our first year we applied a thin top-dressing of compost every time to ensure good germination in the clay soil and to suppress weeds, a year down the track and we have radically reduced our need

So while in our first year we applied a thin top-dressing of compost every time to ensure good germination in the clay soil and to suppress weeds, a year down the track and we have radically reduced our need for compost as a soil conditioner and weed suppressor.

— Mikey Densham

for compost as a soil conditioner and weed suppressor. And we have almost no weed pressure at all, which has been really crazy for us. By following Paul and Elizabeth's basic principles, no-till in this altered context is still working magic for us. Don't get me wrong, the soil in the other areas is improving as well. Over the whole farm we are seeing the soil strata change, increased worm activity, etc," said Mikey. "However there's been a far more dramatic change in soil workability in the areas with the overheads."

"Some people think, 'Oh, my farm is on drip,' or 'My farm is on overhead.' I like to think in terms of how these tools are changing and working with our soil, and how they are influencing our crops' growth. Overhead and drip can give you very different results. It can actually help develop soil in different ways. Now I think we're going to be using overhead as a way to improve our soil's structure and its overall tilth. In the areas of harder clay we are going to use it to transition the soil faster into a more workable structure," said Mikey.

The Philosophy of No-Till

"We're in the beginning of everything, and for us it's such a great thing to have no-till as a methodology and guide. I feel like I have such a backbone behind what we do, and pushing through with it and not giving up when it's really hard is the biggest thing for me. Maybe in a few years we'll have better tips! But now for me the biggest thing is having that foundation, which we really believe in, and drives our decisions," said Keren.

"The more people that are moving away from relying on tillage—even if they're tilling once a year at the beginning of the season, and they're not tilling between every single crop—is a good thing. That's already going to build their soil, and hopefully make it more productive, which is a hundred percent what it seems both Conor and Paul were saying: that there's an economic backbone to being no-till, not just the ecological side," said Mikey.

"That's a massive advantage we have. I feel like it pushes you for production, because the minute we cut the crop out, we immediately want to put something in [to keep the soil covered], so there's a constant rotation of crops, and there's no need for stale seedbedding. It just pushes production in an amazing way," said Keren.

"We're in a wet area, and you cannot even drive a tractor up our hill in the winter. We're getting crops in and out of the ground in early, early spring because our beds are formed and they're ready to rock and roll early in the season. So that's been a massive one for us, and I think that'd be a tip to share with people as well. You can get ahead in the market by using methods like this. You can get your crops far earlier than other people because you're not relying on heavy machinery," said Mikey.

"The first season we were growing on thirty beds, and the amount of food that came out of there was [very large]," said Keren.

Mikey at a farmers market.

Credit: Mossy Willow Farm

"We went from zero to a hundred in two and a half months," said Mikey.

"And how much land are you farming on now?" I asked.

"Three quarters of an acre," said Keren.

"Is that how much land you're trying to do, or are you still making new beds and expanding them?" I asked.

"We're still going to expand a little bit more. I think we're going to try to hit right about an acre and hold at an acre. I think that seems to be pretty good in terms managing the labor and it's also about crunching the numbers to make sure we can produce as much as we need from it to make a living for ourselves and the farm workers. I reckon an acre's going to be the sweet spot for us," said Mikey.

"Is there anything that I might not have asked you, or that I should have asked you? Or any final thoughts?" I asked.

"I love how when you talk about no-till with

some [people], so much of it has to do with practical technique, money, [quick bed] turnover. And then you speak to some [people] and most of what they're talking about is ecology. It reaches and touches so many elements of farming, not just the practicalities of not tilling. It's an ecological backbone, a philosophical backbone, an economical backbone. It's something that has a philosophy. I think no-till comes with a vast package that expands well beyond just not tilling," said Mikey.

"Yes, that's really why I'm so excited about no-till. The exciting thing for me is that they're both there—the ecological and economic sustainability," I said.

"But I do think it will take a bit of time. For us, for example, there is an economical advantage because the turnover is really quick, and we get a lot out of the ground, but at the moment our soil profile is very challenging. It does make certain tasks on the farm take way longer. Because we're on such hard clay," said Keren.

"In Australia, the market gardening scene is still small and tillage is a mainstream practice," said Mikey. "Where we are there are not many veggie growers. Because of the way we farm we are able to farm where most people simply wouldn't grow veggies. Because of the 'lightness' of our farming practices upon the landscape we've managed to be very successful in this untraditional location. And we have vineyards and high-class restaurants all around us. We're cherry-picking all the fine-dining restaurants to sell to, where I think if you had a tractor, no one would even think about doing veggies where we are. They'd be running up and down the hill in a tractor, slipping and sliding. They just wouldn't do it. So, definitely a part of our success has been choosing the spot we're in and practicing how we're farming because, yep, you just wouldn't do it otherwise," said Mikey.

"I do think there's potential for people to use the fact that they're not tilling as part of their marketing. I don't know if it's something that will gain momentum over time, because I think right now, if you said 'no-till,' a lot of people wouldn't know what you mean. People are so disconnected from agriculture, most people don't think about the mechanics," I said.

"I agree that on the sign, it's not a buzzword that people recognize or understand. And I think that is ridiculous. Even for me, when we came to Singing Frogs after having farmed on our own and hearing that tillage releases nitrogen and carbon from the soil, that turns into carbon dioxide and nitrous oxide, which are two potent greenhouse gases. That was alarming, scary, and yet exciting, and emotional. People don't know that tillage is such a massive environmental hazard. Even people that are in this game [of agriculture]. That really drives me forward, trying to share that. It just feels insane that it's not a known piece of information," said Keren.

On my way to Natick Community Farm, I thought perhaps I had gotten lost. Driving in from the west, gradually there was less and less open space and more suburbia. It didn't look like farm country. But then I hung a right and there was a beautiful little farm in the middle of it all.

Historically the farm, which is just off the Charles River in Natick, Massachusetts, grew flowers for sale at the Boston Public Market about twenty miles away. Now the town of Natick owns the farm, which is run as a nonprofit organization for both production and education. Being at the edge of the Boston metropolitan area, the farm offers a variety of programs for young people. The farm sells their seedlings, vegetables, and flowers through an on-farm store, CSA, and wholesale.

NATICK COMMUNITY FARM

Casey Townsend and Dan Morris
Natick, Massachusetts
Mixed vegetables and flowers
Solarization, compost and leaf mulch

I met with assistant director Casey Townsend and vegetable grower Dan Morris, who showed me around on a hot summer day following a cool spring.

One of the most interesting things about doing these interviews was how much growers referenced each other. In this case, Casey got the idea to try no-till from Connecticut grower Bryan O'Hara (see interview p. 305). Casey saw one of Bryan's presentations and decided to try it. Having just interviewed Bryan a few days prior to my visit with Casey, it was interesting to see how Bryan's system translated to another farm.

"The first summer, three years ago, we did a hundred feet of no-till beets next to a hundred feet of beets that we tilled. And the no-till beets turned out a lot better, for several reasons: one, those beets demonstrated higher brix values; two, we have seen better coverage across the beds, so we are using the entire bed space and are therefore seeing higher profits from those beds; and three, they had fewer pest issues. So the next year, we did one 100' by 75' quadrant of no-till. Just from the results we saw that first year, we thought, 'this makes sense.' So we piloted it that first year and from that, we've just been expanding every single year," said Casey.

What Casey noticed about those beets and subsequent direct-seeded crops is that "the seeds were less concentrated in rows and therefore did not require any type of thinning. Also, we sowed the seeds at a rate so the entire bed was covered with beets (in comparison to single rows). This creates a more productive bed financially, and in the sense that the beet leaves tend to shade out germinating weeds. So it helps us harvest more from the same space while also putting less labor cost into each bed. Covering the entire bed also helps manage the microbial life in the soil as well, as there tends to be less 'static' soil with no plant life feeding it."

The seed-starting greenhouse on Natick Community Farm.

The System

At Natick Community Farm, they use Bryan O'Hara's solarization-based system. "We don't usually rely on occultation because it is such a long process (4–6 weeks). So our normal process is: 1, mowing any weeds and remaining crops down; 2, solarization; 3, spreading compost; 4, planting seeds; and 5, covering with a mulch barrier (usually leaves). We use occultation for the sides of our beds to keep quackgrass out but only when needed," said Casey.

Between crops, they use solarization for a few days to kill any remaining weeds from the previous crop. Then they put down a layer of compost, and seed into that. They have found that pulling a leaf rake upside down over the seeds makes enough seed-to-soil contact to get good germination.

To transition from a cover crop, "We mow it, then we solarize. We always add a little bit of compost. If it's a new field, we add an inch to two. If it's an existing no-till field, we try to add half an inch. Then we broadcast our seeds into the bed," says Casey. They're playing with whether they have to add a mulch layer on top of the seeds or not.

Whereas Bryan O'Hara adds a layer of straw, wood chip, and leaf mulch to the top of the soil, "What we do is; because leaves are free, we top-dress with leaves. We have a leaf chipper, but it is super labor intensive. So usually I have my summer crew just make a pile of chipped leaves," said Casey.

"I did use his drag chain for awhile. And I found that I didn't like it. We stopped using the drag chain because we felt it was moving the seeds too far down the bed. With the rakes we are able to better control the movement of seeds. So we just use the backside of a leaf rake and pull it over the bed. Then we apply that little bit of leaf mulch on top, and water it in. Then we usually have to weed one time. So basically Bryan's system is what we're doing and we're trying to figure out if we can get away without the moisture barrier [of leaves on top].

"Even better we apply the compost, add the seeds, put the mulch on and it rains. That's the ideal situation for us. Because all that stuff is living matter as it comes out of the compost, you've got all these microbes that are working in there. So if you expose them to the sun, as Bryan [O'Hara] says, you're going to kill them," said Casey.

"With the back of the rake you're really not moving much around. You're letting the seeds just [settle] into the mulch. Our leaf chipper stopped working and it was so wet that we stopped making leaf chips," Casey said. "So we started just raking a little bit more to see if that was as effective, and it's seeming like it is as effective. So we might modify that part, [and just leave off the mulch]."

Mulching

This temporary strip of landscape fabric is knocking the grass back from the border of the field.

We walk over to a block of salad mix beds to look at the system in action. "We took a big gamble here. We didn't have time for [the leaf mulch], and the leaves were too wet. That's another leaf problem. When the leaves are too wet to go through the chipper. So we just broadcast [seeds] and skipped [the mulch]," said Casey.

Credit: Andrew Mefferd

"You know what's interesting, now that I think about it? The spinach seeds are about the size of a beet seed, right? And those worked," said Dan.

"I think it's going to come down to figuring out what seed size needs to be covered and what doesn't," said Casey. "The cool thing about it is you can seed with your hand. It's just like tilled soil all the time. And the best thing is when you start digging down here, you'll see all these fungal mycelia. And in the hoophouse, it's even more evident. You see fungus [in the soil] all over.

"One of the downsides of using a leaf mulch is that I never had problems with squirrels or chipmunks before. But you'll get acorns in the compost, and you'll be walking through going, what is that hole? And the chipmunks will go in and try to find the acorns, and they pull them out and make little holes in everything. That's one of the weird things about using leaves," said Casey.

Crop Rotations

"In 2015 we added the second no-till block. We just keep adding blocks as we make more compost. Roots are over there now, alliums are here, and then this is going to be our lettuce block next, and we just keep rotating things around as per our certifier," said Casey.

"We rotate the garlic area every year. So the mulch from last year, instead of tilling that in, when we pull the garlic, we leave the holes, plant our kale into it, and then we'll use that for next year's mulch. I call the guy down the street, and say, 'Give me the gnarliest hay you've got.' And he gives it to me at a really cheap price. And he always looks at me crazily, like, 'What you want that stuff for?'" said Casey.

"No animal should be eating it," said Dan.

And that gnarly hay makes great garlic mulch.

The Foundation: Compost

Since one of the underpinnings of Bryan O'Hara's system is high-quality compost, Casey started making compost specifically for use in the no-till system. To explain their system, Casey and Dan take me over to their compost-making area. They have a large pad with three sections where they can pile and mix ingredients with a tractor bucket.

"I was never really great at making compost to begin with. And I heard Bryan talk, and our system is pretty much based on his. He talked about making this carbon-rich, leafy stuff. We've gotten a lot better at it. This is where our whole system begins," says Casey as he takes a handful of dark-brown compost. "So this is a 25-to-1 [carbon to nitrogen ratio compost]."

"We only use stuff that's free. Because we're in the [Boston] metropolitan area, everybody wants to get

A winter shot of greens in the no-till hoophouse.

Credit: Natick Community Farm

Credit: Andrew Mefferd

Casey Townsend with a handful of the compost he makes.

leaves off their lawns every single year. So I have to turn away landscapers bringing leaves to the farm," said Casey.

"Our system is ten parts leaves, six parts wood chips, and one part manure. The wood chips come from [a landscaping] company, or we get them from the town of Natick. The manure comes from our farm.

"So those are the main components of our compost, because they're free. Leaves, wood chips, and manure. That's what we base everything around. It took me a year before I figured out that I need stuff on site all the time.

"Because we're organic, for our certification, we have to prove this gets up to 130 degrees F for four days. That's not a problem. Usually it's up around 160. Bryan wants it lower so you don't kill the microbes. So it's a fine balancing act to try to keep it closer to 130 than 160 degrees," said Casey. "But this is the real labor-intensive part of no-till. Everything else is basically harvesting after this."

Though they have a few tractors, the no-till is mostly done by hand. They have a tractor with a bucket used to turn the compost, a Farmall Cub to cultivate the tilled part of the farm, and a BCS walk-behind, which does the rototilling. As far as the labor-intensive part, unfortunately, their compost-making area is on the other side of the farm from the fields. So, when no-till beds are prepped, they spend a lot of time moving and spreading compost.

One of the benefits of the no-tilled part of the farm is that they no longer have to prepare as many transplants, on such a rigorous greenhouse schedule, because they can replant successions so quickly. With solarization, they can have a crop removed and re-seeded within one to two days.

"One thing I took from Singing Frogs [Farm in California], is they always have tons of transplants on hand. We always want to have something ready to go into the ground. So we just have these," said Casey, showing me a bench full of transplants. "We do lots of greens and lettuce mix. The lettuce mix area I want to show you is really amazing. No-till has changed the way we do salad greens."

"We grow extra trays just in case space opens up, so we have something to throw in the ground. Usually lettuce is that extra thing," said Dan.

"We have a standing order with the school for lettuce mix, which around here is pretty pricey stuff, $9 to $10 a pound. So we're just cutting [lettuce] all the time," said Casey.

The no-till hoophouse in summer.

Direct Seeding

There are some aspects of this particular no-till system that are a departure from the way most people grow vegetables. Broadcasting seeds on the bed vs. using a seeder, for example. This no-till system is one of the only ones I can think of that uses broadcast seeding for vegetables like carrots. Most vegetable growers are used to using a seeder when direct seeding most crops.

"Bryan talks about how it's always easier to teach people to use no-till who've never driven a tractor before. And I find that's the case. People who know how to use tractors, they're just like, 'What are you doing there?'" said Casey.

"It's an unlearning first before relearning," said Dan.

"Yeah, broadcasting is always hard. We've narrowed it down to only a few people who are allowed to do it, and then we just have to irrigate," said Casey.

Credit: Andrew Mefferd

Solarization and a Little Bit of Occultation

In the past, they were moving around multiple pieces of clear plastic to solarize a large block. But they just recovered a greenhouse, so they kept the old plastic and can now cover an entire 100' by 75' block, or an entire greenhouse, with one piece of plastic.

"Even though we solarized, we're still dealing with stubborn perennial weeds," said Casey. "Bindweed is in this field, and in [the greenhouse]. We closed the greenhouse down and covered the soil with clear plastic to solarize. But you still have to deal with perennial stuff all the time."

"In the field, we solarize for two days usually. It gets really, really hot underneath the plastic, and then we pull it off. And it's always a fun team building activity to try to move the big tarp. It's a beast."

The summer no-till hoophouse. Tomatoes to the left and right are interplanted with basil and carrots, which will be harvested by the time the tomatoes are yielding.

Credit: Andrew Mefferd

As far as solarizing in the hoophouse, they can do it later in the year than they can in the field, because the hoophouse is so much warmer than in the field. "During the summer we grow tomatoes [in the hoophouse], and we push everything as hard as we can. And then we solarize in November." The late fall solarization wipes the slate clean before planting winter crops.

"One of the other problems we are finding with no-till is, because you're not tilling, you get this edge effect every single year of the grass creeping in. So we just move this [black tarp], for occultation. Jean-Martin Fortier talks about it a lot. So we just move this around, depending on where we want to make new fields," said Casey. "And we just moved this one out after four to six weeks. So what we'll do is we'll probably apply compost into this, now that it's been tarped [to smother the weeds]."

Weeding

"In terms of weeding, we front load the work. We put all that compost into it, and then that's our workload," explains Casey, gesturing towards a bed of recently cut salad mix. "I walked ahead of whomever cut [the salad mix] for maybe two minutes and weeded it, and that's all the weeding we will do for this entire block before we harvest it. It's amazing."

Dan explains how they really notice it in August, when it's all they can do to keep up with the harvest. Not having to weed helps keep the busiest time of year manageable, instead of having to harvest and weed at the same time.

"So is that accurate to say you're going to put two minutes of weeding into this 100-foot bed?" I ask.

"Let's say three minutes. Just to be on the safe side," Casey laughs.

"One thing we always do is, we're still trying to play with the numbers. So we put a stake in each bed with the day, we write what was planted, and we write the number or the ounces that were seeded there. So that way, if a bed doesn't work, [we can try and figure out] what happened," said Casey.

"And [the variables] could be one, irrigation, it could be two, who was the broadcaster, or it could be three, the compost. And I'll show you some of the crappy compost I made that I jumped the gun on. I was like, 'Oh, this is gonna work,' and then it didn't work at all. So, you get better at making compost. You get really, really good at it."

Looking at the no-till beds, there are some weeds poking through the crops. "As you can see, yellow dock is out there. Buckhorn plantain. You still get these perennial weeds. Bryan says to cut them out, so that's why we always have these knives on us. But I found with some of the stuff, you just gotta pull the roots out," said Casey.

Credit: Andrew Mefferd

These greens were planted by scattering the seed by hand.

Credit: Natick Community Farm

Solarizing a bed on Natick Community Farm.

No-Till Later Than Tilled Ground

Standing in the hoophouse, Casey tells me that ground that is not tilled stays cooler for longer in the spring than tilled ground. Which is especially noticeable in a cool spring like this one.

"The other drawback of no-till is that, in the field, it's always a month behind, because you're not tilling it and getting it warm. But then last year, for example, we were pulling stuff out of no-till until November. It's always behind. It's just slower to heat up. So our plan is to do more overwintered stuff in the no-till this year, to see how that works."

"Yes, leeks and parsnips survive the winter here pretty well," said Dan.

I MISSED THE TURN INTO NEVERSINK FARM A COUPLE of times. Though it's only 120 miles north of New York City, the Catskill Mountain terrain is steep and it's surrounded by state parks, nature preserves, and national forests. It feels about as far removed from the city as possible, like Rip Van Winkle could still be napping around here somewhere.

Conor Crickmore
Claryville, New York
Mixed vegetables
Occultation

The farm is tucked in between the road and the Neversink River. Looking back on our interview, I realize one of the things that makes Conor's farm unique is that he has cut everything out of his operations that doesn't need to be there, and reduced farming down to only what it absolutely needs to be.

There are three things I really like about Conor's approach: he's cutting down to the essential; he's not afraid to talk about money; and he embraces very small farm footprint. Conor realized he was wasting a lot of time and effort on tillage, and that it could be cut out, and so he did, which is one of the things that leads him to make the statement on his website, "Our farming practices may be radical but they have resulted in our farm being one of the highest production farms per square foot in the country."

As for money, it's really helpful when farmers talk about how much they make as a benchmark for other growers. We are not going to have a thriving small-farm economy without small farms making money. We won't get more of the world's food coming from local sources until we have successful small farms staying in business.

As the editor of *Growing for Market*, I know that Conor raised some eyebrows when he said in an article that, after five years, he was "able to come very close to our early goal of 400K on just 1.5 acres. I now feel that for us, 1.5 acres is the comfortable limit if we wish to maintain a healthy production-to-profit ratio while also having a limited staff and working reasonable hours."

Before running that story, I double-checked the number with Conor to make sure it wasn't a typo, because it's so much higher than many people expect. Having been to his farm, I don't doubt him. Walking around his small, intensely planted farm reminded me of a greenhouse. Even in the field, the way plants are crammed in as tightly as possible, with all growing space always producing, is the greenhouse mentality writ large. When I talked to him, Conor revealed that one of the inspirations for his system was taking greenhouse techniques into the field, and everything really started to make sense.

The greenhouse is a high input/high output system. Greenhouse growers typically put much more labor and energy into, say, a tomato plant than do field growers, and are rewarded with much higher yields. This is the mentality that Conor has put into his growing outside as well as inside.

People would have been less skeptical of Conor's numbers if they had known how many people he has working for him: four (or more), to help him maintain that acre and a half. And a significant portion of that acre and a half is in greenhouse production.

A summer field on Neversink Farm.

Credit: Neversink Farm

Yet another important point to make in understanding Conor's numbers is that he sells much of his produce at farmers markets—so he's getting the full retail price, minus the investment of labor in travel to and staffing the stand. Conor's gross is still significantly higher than average, which is one of the reasons I wanted to talk to him. And that makes sense. Neversink Farm has chosen intensive production over extensive production.

A way to break it down is to compare Neversink to another farm I visited—Singing Frogs in California. For example, Paul and Elizabeth told me that their system required 1.5 people per acre for them to make $100K per acre off three acres. Conor has doubled the ratio of labor to land, and the corresponding output from that land. Instead of having 1.5 people per acre, he has close to 1.5 people per *half* acre. And by doubling his investment in labor/area, he grosses about twice as much as Singing Frogs does by area. It's the old distinction of growing intensively vs. growing extensively.

Another big difference in the farms is climatic. Neversink is in the Northeast, so they need to make more extensive use of greenhouses and protected space to meet their goals. Singing Frogs, in California, make use of their relatively temperate climate by using a little more space to grow more extensively outdoors. The farms are using slightly different no-till methods to achieve the same goal, of being really productive of both food and income.

Neversink is striving for sustainability on a number of different levels. On his website, Conor says, "We are redefining what sustainability means to our farm all the time. We started with the premise that the farm must sustain our family financially. Within that framework we try to employ more sustainable practices as our farm evolves and grows. Some practices we think add to our sustainability presently: [being] certified organic, small size ... with high production per square foot, no tractors, no plastic mulch, no wax based produce boxes, no hydroponic growing, pay[ing] our staff a reasonable wage."

Conor wholeheartedly embraces a small farm footprint. His farm is a manifestation of the goals he set for himself starting out. As he put it in a *Growing for Market* article, "We wanted to stay very small, but we also wanted to live well. We were not young and since we had no retirement money, we concluded, however naively, that to live well, have kids, and be able to retire at some point, we should try to make at least $400K gross within a few years, and we would expand to whatever amount of acreage that demanded. We were obviously not yet schooled in small scale farming financials, because if we were, we might never have left the city with that kind of goal."

When farms need more income, the first thought of many growers is to get bigger. What Neversink did instead was to become more intensive. This is really important, because small farms will not be able to compete with big farms on economy of scale; they need to find an edge to be competitive with industrial food, by competing on intensivity of production, not extensivity.

With a one-and-a-half-acre farm footprint, Conor is actually shrinking his farm as it becomes more productive. A lot of people with excess production would go looking for new markets, but Conor realized that he didn't want to take on more markets. Learning when you've reached the "right size," and then stopping, is an important lesson.

Inside Neversink Farm's no-till greenhouse.

Credit: Neversink Farm

Many small growers, even latently, think they are somehow less important than bigger growers. Many of the growers I talked to said, "We just have a couple acres," or "I just want you to know we're just a small farm before you come all the way out here." I think this is a hangover from the Earl Butz, "Get big or get out" approach to agriculture. I think that we would be much better off with tens of thousands of few-acre farms, instead of a few big farms. As long as you're staying in business, supporting yourself and feeding other people, it doesn't matter what size your farm is. Thanks to Conor's acre and a half, he and four other people have jobs, and he's feeding a lot of people.

Farming more traditionally with a rototiller caused Conor to think of a lot of ways he could streamline things, which has led to a farming style that is different from most vegetable growers. He has completely stopped rototilling. His thought process has led from, why till if I don't have to, to, why take the irrigation out of the field if I don't have to, to, why be any bigger than I have to? See Conor's website, neversinkfarm. com, and the Resources section for more details.

Motivation

"You've figured out how to be very productive," I said to Conor. "Is that your primary motivation?"

"Initially it was about being more efficient. I would never start doing something and be like, oh, it's going to be more work. That would be ridiculous. I have no interest in making more work. If it's heavier, I don't want to deal with it. If it takes more time, if it's harder to get a worker to do it, I don't want to do it. And, it's a lot easier to show someone how to use a broadfork than it is show them how to use a [rototiller]. Right? I mean, any monkey can use a broadfork. But a [rototiller] is a lot harder," said Conor.

"There are just better ways to spend the money [than on machinery]. It's more efficient, and you do get a slight reduction in weeds. I think the reduction people talk about is maybe a bit overblown, because most of your weeds are probably going to be blown in over time. You need a weed control system. No-till is not going to help you much if you have a lot of weeds. But, if you have a weed reduction system by cultivating and everything else, it's going to probably add in a little bit. At least at the beginning when you're trying to reduce the weed seed bank, you're not bringing them up all the time from down below.

"You can have permanent beds, which is incredibly nice. You can have stakes there, which now you can run a string whenever you plant, and that's going to be a more efficient bed usage, which is really important on small scale. Raised beds are a complete waste of space. You have to maintain more area, and no significant production increase. You know, you need about 30 percent more area for a raised bed.

Credit: Neversink Farm

Transplants along with direct-seeded crops.

"But when you do [flat beds, you can have a string to mark the beds]. The irrigation can be permanent, left there all winter. It never moves. So, in the spring I just need to turn it on. And that's a whole other thing you have to drag in and out when you're tilling.

"Also, you're going to use a little less fertilizer. Which is probably a bigger reason to do no-till than weeds. Because every time you till, it's just like turning a compost pile. Because in organic [systems] you're trying to break down material for fertility. And every time you churn it up, it's burning it up. So we just have a very thin layer at the top, where everything's breaking down, and so you need to add less. And for me, that's good, because it's very sandy soil, so my fertility goes away very quickly.

"And, as I build my soil, I don't keep pushing it down farther, so I don't have to water as much. Because as you go deeper, you'll find the soil changes and gets sandier. So, it's going to hold a little bit more water than if I keep bringing the sandy stuff back up."

"Right, because you're just Tilthing the top, and then adding your fertilizer and compost or whatever and just working the top little bit of soil," I said.

"It's like three quarters of an inch. That's it, I just come over the top, to protect the amendments from going somewhere, because that stuff is expensive and it just speeds it up a bit. Because it's going to take three months for that stuff to break down. Why add a couple of months [to the breakdown] when you can just put it on the top?

"But in the houses, that's not true. We have a better-protected environment, you can just throw it on top and the worms will come and get it. And you don't need to rake it in or Tilth it in."

"So, you just apply your amendments on top in the hoophouses and don't scratch them in or anything?" I asked.

"Not all the time, no. And I would never scratch it in. You know, I may Tilth if I need a nice seedbed. But, if we're just transplanting, it's not going to make any difference. There's no reason to Tilth it. If you can see inside the house, you'll see where the amendments are laid down, and then you'll see all the worm castings. Because at night, if you come out there, there are worms in there that are just eating it. It's just feeding them. And they'll pull it down [into the soil] and you don't need to do anything. Because I find that once you stop inverting the layers of soil, then they just get filled with worms. Especially inside of the house," said Conor.

Inspiration

"When you were starting out, where did you get the inspiration for your system from?" I asked.

"I don't know if it was inspiration. It was more like little steps," said Conor. "We already had a broadfork. It was more what we were doing inside the houses. You can't bring the rototiller in the houses. Especially in the winter. And so, we would just fork it and then Tilth it. So I thought, if it works in there, why can't I just do it out here? What's the difference?

"It was just what we were doing inside. Because I found that inside the soil is better, you can take care of it better. So, I thought, why don't we try to do that out here? We were staking it in there, first, because it's even worse if you need to redo your rows inside of a house. It was just a waste of time. So, we'd stake it out, because you know how many rows there are going to be. And I thought, why don't we just stake it outside too? Because it's such a big problem. As the paths get bigger, people are stepping on the germinated carrots on the edge of the bed. At least if you have some stakes it'll help stop that," said Conor.

"We're putting up a new house this fall, just for tomatoes. And we're going to go to five rows in a thirty-foot-wide house. It may not be the most efficient when you go to greens, but you know what? I'm so lazy that I'd rather grow greens in fewer rows, because I'm not struggling to make money like I was. At the beginning, we were like, 'Oh maybe I should even seed the paths.' I'm not going to deal with it, I just set it up once and am done with it."

"Yeah, there's a lot to be said for just keeping things simple. So, once you've established these beds, do you send crew through on any kind of a regular basis to look for weeds that have, like you said, blown in?" I asked.

"They're always blowing in, so you have to cultivate all spring. And that's when we hit it really, really hard is in the spring. It would start as soon as the snow melts. We're probably cultivating once a week until July. Then, you're just harvesting. So you don't get to do as much cultivating anymore, but that's okay, as long as you got it in a good state. You're not going to get ahead of it, and then we rely on turning over the beds. And once you turn it over, you're taking out any weeds that did get in there," said Conor.

"The most important thing is nothing is going to seed. That's going to be worse than what blows in. If you have a big weed plant with thousands of seeds, it's going to be bad. But we're trying to move more and more indoors, just because it's so much cheaper to grow indoors.

Winter greens harvesting.

Credit: Neversink Farm

"You know, it's anywhere from three to ten times more productive and the maintenance is almost zero. Our houses, most of them we don't cultivate at all. There are really no weeds, where are they going to come in from? If you can get rid of weeds, now you're just printing money. Right? That's where all the money gets chewed up. Either you're going to cultivate them out, which is less money but it's still money. Or they're going to interfere with everything you're doing, which is even more costly."

Markets

"So, what are your markets? It feels like you're out here in the middle of nowhere, but I realize you're not really that far away from New York City," I said.

"Well, we only started selling in New York City this year [2017]. We sold to restaurants in New York City for a few years, but that's not the best money, farmers markets are better," said Conor.

"We started with farmers markets locally, and then we've gone farther out as the production went up. So, for the past few years, we've been doing farmers markets in Westchester, where I grew up. That's why we picked Westchester, because I know it really well. Also, the population density's good enough that you can get some pretty good markets. If you get into the better markets, you can earn enough to make it worthwhile," said Conor.

"It's a five-hour round trip for us, but it makes sense. Locally, we kind of grew out of it. We would have needed ten markets, or more. And if you want to take time off, that's not going to work."

Farm Size

"So how much space are you growing on here?" I asked.

"This year, it's a little less than one and a half acres, because we've been shutting it down, [taking ground out of production] by putting plastic on. And this isn't like the [Jean-Martin Fortier method] where you're doing weed suppression. We put it out of commission. But we don't want to let it turn into weeds, just in case we need it again. You never want to be like, 'Well, we're just going to let it go.' It's there if we need it. The more greenhouses we put up, the more plastic we put down," said Conor.

"We'll just do blueberries or something. But, as of now, we're getting smaller. We're going to look at getting down to 1.3 acres or less next year. We find there's just no reason for it. With the amount we're producing now, we need another farmers market, and we don't want another farmers market," said Conor.

Soil Fertility

"What are you using for amendments between bed flips?" I asked.

"It all depends on what our soil tests have been saying, it depends on the time of year, what was before it. So we have pretty much everything on standby in the amendment room. Anything from feather meal to chicken manure to alfalfa meal, soybean meal, you name it. Because they're all different, and this way we can make a mix if we need to, that covers what we need," said Conor.

"And we need different things for outside than we use in the house. We don't use any of the animal stuff in the house, we just keep it veg. We also have seaweed. I like to stockpile lots of stuff. It doesn't go bad, and it's there when you need it. And if we have something weird going on, then we have what we need on hand to fix it. That feels good, compared to the early years, when we were like, 'Maybe we should just get a 25-pound bag.' Now that we have the money, we just get all of it. If there's a sale, I just get a pallet. It's a big barn. I don't need any room for tractors."

Labor

"How many people do you have helping you maintain the 1.5, or 1.3, acres, whatever you have going on here?" I asked.

"We have two people in production, and then two in fulfillment. I don't like more than three in production, it's too many. We may have four in October, just because I have two really good people coming back who were here before. Having a lot of people in the fall is a great thing. It just helps your spring, because I like to have every bed ready to go in the spring. It's very nice. Not a single weed in the field, ready to go. We do our soil tests now [in the summer], and then we'll fix it in the fall and then [in the spring], there might be a little bit of forking and Tilthing, and we're ready to plant. Which sometimes is a bad thing, because you get itchy fingers, especially with no frost," said Conor.

"One or two nice days, and you want to plant everything," I said.

"Yes, that's what happened with the parsley. Such a cold spring, it all bolted," said Conor.

"With no-till you can just get right on the beds when you decide it's time, whether or not the weather has made its mind up," I said.

"It's really about efficiency; it's really about making the farm easier," said Conor.

"What do you think about the scaling-up of these systems?" I asked.

"We started [the farm with] $30 K, and about eight grand of that was our tiller. We would have been better off with a paper pot transplanter, and a Tilther, but we didn't know," said Conor.

"If you're not making a profit, you're not going to stay in it for the long run. If you're killing yourself to barely make any money, you're going to lose your health. People say, well it's not about the money. Well, it's not, but you have to be making it to stay in business," I said.

"It's absolutely about the money. Everybody's got to pay for their health, everybody's got to live under a roof, we're talking about the basics of life," said Conor. "Money has to come, or there is no farm. There is no farm without some profit."

Broadforking and prepping permanent beds.

"Yeah, well that's why *Growing for Market* is a farming magazine, it's not a gardening magazine. It's about people trying to make a livelihood. I'm trying to help small farmers by helping them be profitable. One of the changes I want to see in the world is to have more food coming from local and small farms. So, to make that change, local farmers have to be making a profit. That's the appeal. That's why I like your writing and what you're sharing with people, because you're sharing good ideas to help people be more successful," I said.

"It's not just about doing it my way, but about how to figure out what's making money, and what's not, and getting rid of it," said Conor.

"Because that's one of the reasons we're successful. It's not because we have a method that's successful, it's because we were very quick to get rid of things that didn't work. You know, we mulched. But immediately, we thought, 'This is not the way to get rid of weeds.' Weed seeds don't disappear because you put mulch on them. And I've got to put down the mulch, I've got to take up the mulch, I've got to be able to seed. So, we went through a lot of different things to cut out the things that didn't make money. If there's something that's not making money, it comes out. We don't shed a tear over it," said Conor.

"But even more important than *how to* is *why*. There's so much information. Like, I don't wanna pick on the raised beds. But I probably got 700 feet of rows with raised beds, until I was like, what am I doing? Why am I doing this? Let me think about this.

"It's just that things have to be done very differently on a scale like this. I don't think you can transfer this to a large-scale farm. You couldn't do this on ten acres, not without a huge capital infusion. It's cheap on this scale. But once you start getting to a larger scale, not so much.

"It depends on your definition of what those things are, right? Everybody's definition of small and big is different. Because to me, anything four acres or more is a big farm. And once you get up to ten acres, I think of it as a huge farm. And once you get to fifty, I think of it as a mega farm. And when you start getting into hundreds, that is a gigantic farm," said Conor.

"Well, this is a more human scale. I mean, obviously farming started out smaller, and really the only way you get farms of that size is having machinery or chemicals doing some of the work. So, I guess to scale this

Broadforking
permanent beds.

Credit: Neversink Farm

up, you would have to have people you could really rely on to make the right decisions. Because, yeah, I get what you're saying. You couldn't be out there seeing every bed every day at a certain size anymore," I said.

"Certain aspects definitely, anything can be scaled up. Anybody can come up with something. They could come up with a tractor broadforker that will broadfork each bed. It's just the style of farming that I do, which is something very personal, I couldn't scale up. It's kind of like saying, 'Can you take this restaurant, that makes these wonderful burgers, the way they do it, and then let's do a thousand of them across the country?' You can't. There's something that's going to be lost. It's like, well, why not? Because you can't duplicate certain things, you just can't. You can try, but you're ending up with a completely different product," said Conor.

"Yes, because everybody knows, it's a chain once you got a thousand of them. It's a chain restaurant. It's not your little burger joint anymore," I said.

"And it just makes more sense. Because there's a lot more people who want to own a farm than want to work on a farm. So it's better to have a lot of small farms," said Conor.

"Yeah. Well, I think that's part of the problem. When I was in high school, my guidance counselor didn't say, 'Oh, maybe you'd be a really great organic farmer.' Almost everything they were suggesting resulted in me being in a cubicle. And that's not for everyone," I said.

"That's why we're putting up another greenhouse, and shutting down part of the field. Because that's just so much easier, every square foot becomes easier. Just make it easier and easier. Why not? I don't want to add and add and add. It's enough money for us, so we're perfectly happy with it. But we just want to continue to make it easier. You know, production at some point just starts coming, as you start fixing things. After a while, it's not like you're fixing everything anymore, like you were at the beginning. So you can just work on making it easier," Conor said.

Here are two articles by Conor that were originally published in *Growing for Market*.

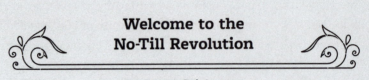

Welcome to the No-Till Revolution

By Conor Crickmore

From the March 2017 issue of *Growing for Market*

I began Neversink Farm six years ago. Building the farm that first year was incredibly hard work, but it was that early experience of struggle that shaped the foundations of how I would farm going forward. My wife and I managed to pull together $30K. With that, a dream, a lot of enthusiasm and not much else, we said goodbye to city life.

We leased a beautiful piece of land just a few hundred yards up the river from a cabin I had purchased years ago. The land had more rocks than dirt, no water, a barn and loads of quackgrass. Our initial purchases for the farm that first year were a small propagation house to start seedlings, a small simple hoophouse in which to grow tomatoes, and a two-wheel tractor to till and plow the fields. We had a great desire to farm but lacked experience and knowledge in the craft.

We wanted to stay very small, but we also wanted to live well. We were not young and since we had no retirement money, we concluded, however naively, that to live well, have kids, and be able to retire at some point, we should try to make at least $400K gross within a few years, and we would expand to whatever amount of acreage that demanded. We were obviously not yet schooled in small scale farming financials, because if we were, we might never have left the city with that kind of goal.

We began our farm adventure with a lot of tilling. Turning a field of grass and weeds into a productive farm can seem like an impossible task. It can be the hardest part, physically, of starting a new farm with or without a tractor. This is where putting the hard work in at the beginning really pays off more than anywhere else on the farm, especially if you choose not to till.

I looked out over our field after we signed our first lease and imagined neat rows of perfect and identical vegetables, where every row is straight and true interrupted only by even more perfect grass paths where handcarts would be rolling around overflowing with produce. In my mental image there wasn't a weed, diseased plant or pest anywhere, but I had no idea how to get there. I concluded that tilling everything in with a walk behind tractor was the obvious choice. I didn't consider any other course of action.

Much of our startup capital was spent on that BCS walk behind, and it was that horsepower that gave me confidence. From sunup to well past sundown, I would till and plow, the tractor bouncing wildly over the river stones. My wife Kate would run alongside to grab the rocks and put them in a cart that we would haul together hundreds of yards to the woods. We made hundreds of trips.

This went on for week after grueling week, tilling sections over and over to try and break up the mounds of grass but they wouldn't disappear. These clods of perennial grass roots would get in the way of everything we did, making us curse as we tried to lay out my imagined straight rows. Planting was torture, as I would hit clumps of grass with the trowel. Forget about using a seeder under these conditions.

In desperation I used plastic mulch in some of the worst areas to try and tame the grass that was emerging at a really scary rate after spring rains. One could not tell where the grass paths ended and our production beds began. There is no way to earn money this way, we thought. What are we doing? We felt Neversink may be sunk. But if either of us seriously felt like giving in, the other would feign confidence that we would figure this out and be successful. We lifted each other up and focused on getting any vegetables we could out of the weed jungle and getting them sold. From that tiny first field of weeds we did earn a small profit, but only because of our free labor. I felt like a failure at farming but we wouldn't give up.

Well that was it. There would be no more weeds at Neversink Farm. That next season, we decided to get on our knees and pull out every weed, inch by inch, foot by root, row by back breaking row until all, and I mean all, the perennial weeds and their roots were gone. Our fingers would bleed, but we were not going through the horror of last season. We were going to put the hard work in up front to make farming easier going forward and this would become a cornerstone of how we farm.

On a daily basis I seek ways of making farming easier. If we could make a huge sacrifice for our future selves and for our farm then we would. Whether it was really hard work or purchasing a piece of equipment that we could not afford, if it resulted in simplicity and efficiency, we would do it, because that will ultimately lead to easy, enjoyable farming and profitability. In farming, simplicity can be achieved by stripping away complexity but can also come from investment either financially or through hard work. The hard work of removing all the weeds made everything simpler and easier for years to come.

We had only tamed one small field of perhaps a half an acre and we knew that we needed to grow, but creating new fields by tilling was not going to happen. Pulling a new pasture out by hand would

Permanently staked beds.

Credit: Neversink Farm

have sent us back to the city, screaming. This time we looked for a simpler and cheaper method. Solarization was where we turned. This is usually done with clear plastic to heat soils up to reduce pests and disease, as well as kill weeds.

Our farm is in zone 5a, so when we used clear plastic, it acted more like a greenhouse in early spring and while some weeds were killed others thrived. We abandoned clear plastic and used black. Solarization is time and temperature dependent. You can have a lower temperature under black plastic, but it will need to remain in place longer. A weed seed will cook when flame weeded at scorching high temperatures and be killed in seconds. Lower temperatures for a prolonged period of time will do the trick as well, but that time must be increased drastically. A few seconds turn into months. Since we had the time, this method was perfect.

That spring, I spread a large piece of black plastic on top of a new section of virgin field. The plastic was removed the next spring and underneath was bare soil with no weeds or viable roots. It was amazing. I continued to reuse that piece of plastic, which is the same size as one of our field sections, by sliding it from a finished section over to a new section each season to open up new ground. We were ecstatic at how well it worked. We laid out rows, forked and began planting. The perennial weeds never came back and for a no-till system to be highly productive, it must be free of all perennial weeds. It was that hard work in the first season that solidified my policy of seeking easy solutions wherever possible.

Tilling is generally done at least once a season over the entire farm and then after every turnover of crops. Before tilling, irrigation and anything else in the field must be removed. After tilling, the beds then need to be laid out again and the irrigation brought back in. To be ultra-efficient those beds need to be measured precisely, and made straight so that no space is wasted. The footpaths also must be marked out in some way so they are clearly visible. After all of this hard work, you then need to do it all over again next season. On top of that, weed seeds are being brought to the surface and the soil is getting overworked. Thus tilling also has the side effect of increasing the amount of required cultivation and soil amending. Never tilling again was an easy choice, so I just stopped. I never wanted to be in the weeds again.

Neversink was suddenly a no-till farm. The tiller and the two-wheel tractor, I came to realize, were unnecessary, really hard work, time consuming, and added too much complexity to our systems. There would be no tilling done from here on out and the time-consuming task of laying out rows would be done once. All of our beds are now permanent. The beds are staked and the rows never change size, and they are ready as soon as the snow melts. The bed stakes are there waiting for us to attach guide string for our seeding and planting. The irrigation can be left in place, avoiding the task of dragging it out from the barn each spring.

Each permanent bed has been assigned a unique ID and I created easy to read maps that I use to assign tasks to specific highlighted beds. What could be a more permanent and simple solution on the small-scale farm? One of the biggest misconceptions of no-till farming is that we must be breaking our backs. I think people picture me strapping plows to my workers like draft horses. I stopped tilling or using the two-wheel tractor because it is much easier, more efficient, simpler and more profitable for us. It has made farming at Neversink very enjoyable. It was the best efficiency improvement we made.

Once we were on the road to efficiency, I wanted to maximize it. Thus my goal is to have every square inch of our 1.5 acres producing to its potential and producing all the time. To achieve this, any bed that is harvested in the morning, I want replanted by the afternoon. This is done across the farm. Because we don't have to clear large areas to get a tractor into, we have the luxury of being able to replant or re-seed any sized area.

The permanent beds are prepared or turned over with four easy steps. First, vegetable matter is removed though roots of some crops remain. Depending on the crop this initial step can be done a few different ways and I have put much thought into efficiency improvements for this step. Second, the bed is deeply broadforked. In a field of rocks this can be really hard, but we already put the work in and cleared out any rocks so that forking is a very pleasant activity. Third, the bed is amended and conditioned depending on that bed's needs. Last, I use a Tilther.

The Tilther is my power tool of choice. It is light, simple, and does the job. It disturbs only the top inch of soil and works in the amendments and smooths the bed. I can throw the Tilther over my shoulder and carry it around the farm. It is great in winter since it produces no exhaust to poison me while using it in the greenhouse. My four-year-old can walk alongside me while I use it. They are cheap, so I can have them at easy-to-grab locations around the farm. The Tilther is the elegant simple solution for "Neversink No-Till" farming.

After a couple of seasons of dialing in what we now call "Neversink No-Till" systems, the field started to look as neat and productive as I had imagined it that first year. There were straight rows, no weeds, and healthy uniform vegetables. I no longer felt like I failed at farming but that I had found a way I could be very successful through simple solutions. I don't want to work harder than necessary. That is why

Scallions harvested in the morning and replanted the same day.

Credit: Neversink Farm

I spend a lot of my time looking for the easy and simple solutions. Farming is my dream job and building it is a joy, but the work I love most is the work that results in less work. Farming can be and is extremely complicated. Thus, I enjoy watching my farm move towards successful simplicity.

Using no-till practices that were developed at Neversink along with the important lessons learned early on, we quickly grew the farm over the following years. The farm grew mostly in production but not in size. These important lessons were to never add more work but rather to reduce it. Use permanent solutions and keep things simple while investing in infrastructure and new systems. Field sprinklers are a great example of the results of these lessons. They are a large infrastructure investment. In a no-till system they can be left there permanently and unlike drip, sprinklers irrigate beds evenly and quickly which increases the germination rate of seeds and the survival of seedlings. Drip can be time consuming to manage while our sprinkler system is almost no maintenance. Also sprinklers are not in the way when cultivating as drip can be. When I want water, I only need to lift the handle of the hydrant.

Simple tools used on Neversink Farm.

Credit: Neversink Farm

When deciding on an improvement to the farm, I not only ask, "will this improve production?" but I also want to know if it will increase work. A good example of where we implemented the lesson of not adding work is in high tunnel design. At the beginning we used caterpillar tunnels since they were cheap. While they do increase production slightly, they also brought with them a lot of problems, and a lot of work. They need a lot of attention and only made sense when we wanted to grow our production regardless of the resulting work. We quickly abandoned caterpillar tunnels and moved towards more high-tech, better-built structures that require almost no attention. These structures increase production dramatically. I'd rather have one great no-maintenance tunnel than ten cheap ones. It is better to have beautiful healthy production on one bed than not-so-great production on many. With a better tunnel, I can do more with less. This philosophy is also why we do not use row cover and low tunnels in the field. Their promise of increased production comes with a lot of work as well.

I also found that field plastic adds work and problems. While we did use plastic to open up new fields, we really try to avoid using it or landscape fabric on production beds or next to hoophouses. Voles and mice love it. Instead I put the work in up front to reduce the need for plastic by removing weeds, and building permanent gravel trenches alongside tunnels.

Stripping down our record keeping to the bare essentials was a huge time saver. I started with very complex spreadsheets for everything around the farm and now have what I call "cheat sheets" that give me all the information on one page. My entire year's planting schedule is on one small laminated card. I have simple task tickets to direct employees in their work. I do not keep elaborate records of everything harvested, planted, and sold.

Rather I use an easy "check-in" process to see if a crop or method is profitable. I randomly take a look at a whole system, like the lifecycle of a box of carrots from planting to market to see if carrots are earning money and how much. By doing this intermittently or after an improvement, I can see if we are earning a profit, or if a system is getting worse or improving. I do not need to check every box of carrots but only one or two boxes every couple months. If I am duplicating a profitable process, then I know every box of carrots is profitable. Cheat sheets and check-ins are so easy, so simple and a lot less work.

After reading books on soil, I was just getting more confused, so simplifying it was the only way for me to make sense of it. Soil fertility can be an extremely complicated subject and while I wouldn't claim that the vast knowledge in that field is unnecessary, I do feel that for me I needed to break it down. I simplified my soil building program to three simple steps. The first being pH and calcium balancing which is only done every two years and I keep careful records on this. The second step is soil conditioning which is based only on the feel of the soil. The last, which is done regularly, is fertilizing and is based on soil tests and crop needs.

All of these changes had the cumulative effect of leaving me vastly more time, which I spent managing rather than working the farm; the more time I spent managing the systems the better the farm did. Our numbers started climbing rapidly and as a result I could invest in more small and large improvements. I could monitor our systems for bottlenecks— places where production is squeezed and thus slowed down. I could also slowly organize and systematize many areas of the farm that I had ignored previously. We thought we would need at least four or five acres in production to reach our early revenue goals but our production climbed much faster than our need to expand acreage. By year five, we were able to come very close to our early goal of 400K on just 1.5 acres. I now feel that for us 1.5 acres is the comfortable limit if we wish to maintain a healthy production to profit ratio while also having a limited staff and working reasonable hours.

Our farming start was extremely hard, much harder than it needed to be. But it was through that experience that I was able to develop our no-till systems, our simple solutions and our emphasis on reduced work. We suffered due to our inexperience, but we learned fast and we continued to master our own stripped down and efficient system of farming. It was the suffering that taught us to change and not to be stuck to any one way of doing things and to design every process with simplicity and elegance. I learned to avoid adding more work for small production gains. That early hard work also taught me to put the work in up front if it resulted in less work going forward. We were able to turn failure into an incredibly productive and successful farm.

The number of small farms is rapidly rising. This is a great thing. There is a small farm revolution happening and I now have time to help other farms hopefully avoid the mistakes we made. I exchange information with other farmers through our social media, during our on farm courses, and my consulting work. I know we can do better together and sharing knowledge is the best way for us all to get to easy, enjoyable, profitable farming.

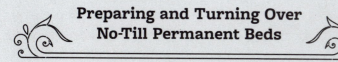

Preparing and Turning Over No-Till Permanent Beds

By Conor Crickmore

Sometimes when turning over beds, I would try to rush through it and not do a very good job. This would cause a whole mess of problems: uneven growth, weeds, beds that are too narrow, too much of the last crop poking through, and an all around reduction in production. Now when I show others how to perform the turnover, I stress the importance of thoroughness. Making sure that the beds are laid out exactly and that turnovers are complete has increased my success rate dramatically and made all bed tasks easier as a result.

In my previous article, I discussed the importance of starting with a planting bed free of both perennial and annual weeds. I also covered how to achieve this in both the short and long term. Once beds are weed free, the next step

From the May 2017 issue of *Growing for Market*

is to lay out the grid and make them permanent. The permanency of the beds is in itself a system of efficiency. Beds don't shrink from foot traffic and irrigation can remain in place. The layout of the farm stays the same from season to season. Beds are ready to grow when the snow melts, and farm maps can be produced with each bed assigned a unique ID.

I use the standard 30" bed. Most tools are designed to fit a 30" bed, so that size makes the most sense for me, since I would rather buy my tools off the shelf than have them made custom. There is more freedom when it comes to path size. I chose 14" because it works well with a 12" wheel hoe blade, which is the widest that is made. Our harvest totes are 13" wide, so to save space, I felt 14" was the tightest I could go while still having paths that are comfortable to work in. A path of that size is easy to stand in and place a bucket on the ground, even for someone with big feet like myself.

All the beds at Neversink Farm are staked at the corners. This helps with their permanency. Without this, it is the path

The tilther doesn't create any fumes and can be used anytime, anywhere even with a four-year-old.

Credit: Neversink Farm

that gets wider over time, reducing the production area. This results in the edges of the beds getting trampled, ruining a large percentage of crops. Thus staking them in some way pays for itself very quickly.

I keep the following tools on hand for staking beds accurately: a hammer, two metal spikes, a bed template made of wood for the placement of stakes, a template for depth of stakes, and a spool of string. I use one-inch square hardwood stakes that are placed at the four corners of every bed. The spool of string does double duty as it is also used as a straight edge between the stakes when planting or seeding.

Laying out a section of field begins by placing the metal spikes at each corner of the field. I then run a string between them at about ten inches off the ground. This serves as the guide for the bed stakes, so that all bed stakes are placed in a straight line. Then using the bed template, I work from one side to the other hammering in the stakes to the desired depth. Depth is checked with the depth template. I use a push spreader on the field, so my desired depth is one that gives the spreader a slight clearance.

The bed stakes are placed in the bed rather than in the path. Thus the 30" is measured from the outside of one stake to the far side of the next bed stake. This enables clearance for the wheel hoe and people's feet. Bed stakes need to be replaced over time, so I keep the templates, a hammer, and stakes in a kit on our tool racks and I make stake replacement a required step during the task of turning over beds.

Now that all the beds are staked, we can move on to prepping or turning them over. Preparing a bed is the same as turning over a bed minus the first step of removing vegetable matter. Bed turnover takes place between the harvest of one crop and planting the next. The following tools are used when the permanent beds are prepared or turned over: a weed hook (like the CobraHead), stirrup hoe, buckets, broadfork, bed stake replacement kit (if necessary), and a tilther. There are four easy steps we always follow.

Step one

All vegetable matter is removed. Root crops like beets, carrots and radishes are the easiest crops to remove since most of the plants should have been removed during harvest. Depending on how they were harvested, there may only be tops to rake up. Larger crops like chard, kale, cukes, and zucchini are pulled by hand. There are usually no more than a plant every foot so I consider these crops almost as easy. Head lettuce, bok choy or Napa cabbage is cut at, or just below ground level during harvest, so the only green matter removed from these beds during a turnover is usually stray weeds. The roots of these crops are left in place.

Baby greens are the most challenging. I have new field hands practice turnovers on baby greens. It takes skill that is acquired over a few weeks to learn the quickest method for the different conditions one could face. Overgrown greens almost always need to be pulled out by hand, which is something to be avoided. Young greens should be cut as low as possible with the greens harvester first. Even if the bed was already harvested, it is good to go at it again and give it a close buzz cut. While doing this, I tilt the harvester back slightly to keep the blade away from the soil. For many greens, like baby kale or arugula for example, this will kill most of the plants since the cut will be well below the growing tip.

After a quick haircut with the harvester, the greens are then cut at ground level with a sharp stirrup hoe, leaving the roots behind. We keep a couple of short-handled 6" stirrup hoes on hand for this purpose. This is done carefully, row by row, so that all plants are killed. The bulk of the leafy matter is taken away while some of the dead leaves and all the roots are left. This technique takes a bit of practice to be able to do quickly. One's technique is important. It is best to remove baby greens immediately after harvest. Younger greens are removed more readily so I recommend not harvesting more than a couple of times prior to turnover.

When I say to remove vegetable matter, I am including weeds, thus the need for the weed hook. It is important to start with a completely clean bed. One should never plant or seed into weeds. If you wish to use the stale seedbed technique after turnover, you will be much more successful if you finish the turnover with no weeds.

Step two

The second step is broadforking. This is done deeply until the fork's crossbar hits the soil. The length of the tines determines how far apart to fork. Thus broadforks with ten-inch tines are used every ten inches along the bed. Removing any rocks that the fork hits is very important. I forbid walking on planting beds and I add amendments that will keep the soil loose, because non-compacted rock-free beds are very pleasant to fork and each time rocks are removed it makes the next turnover easier.

Step three

The bed is amended and conditioned depending on that bed's needs. In an upcoming article, I will discuss our fertility in detail but during every turnover something is added for either fertility or soil condition or both. At this point any bed stake that is rotten or missing would be replaced.

Final step

Lastly, I use the tilther. The tilther is my power tool of choice. It is light, simple, and has just enough power to do the job.

A bed must be free of rocks, most vegetable matter and be forked for the tilther to work well. It disturbs only the top inch of soil and works in the amendments and smooths the bed. Organic amendments will not do much if left on the surface. To see their benefit this season, they must be worked into the soil so that they can begin to break down. That one-inch deep tilthing is all that is needed. Lettuce head roots pop out of the soil with a slight tug back on the tilther. I keep a tilther along with all the other tools mentioned on every tool rack around the farm.

No till production, for me, is quick and easy. Turning over beds rapidly is key to high production. This no-till bed turnover really helps in having every square foot of ground producing, since it can also be done inside greenhouses in winter when you don't want to bring in heavy equipment.

Neversink Farm's No-Till Definition:
We all have heard the adage, "there are as many ways to farm as there are farmers." No-till methods are no different in that there are many types of no-till systems. At Neversink we do not use a tiller, nor do we employ a two-wheel tractor. I have defined tilling at my farm as any inversion of the soil that would change its natural layering, or the regular use of heavy machinery. This is my personal definition and not intended as a label that I use beyond my farm. Outside of my farm I accept far broader definitions of no-till systems as perfectly valid.

I VISITED WITH ELIZABETH AND PAUL KAISER OF Singing Frogs Farm in Sebastopol, California, on October 10, 2017. Historically bad wildfires were burning land, homes, and some of Sonoma County's famous vineyards just miles away. The fires were so close it was hazy when I arrived from the smoke. It was an unusually quiet day on the farm, with the Kaisers sending their crew home due to the poor air quality.

I could see how on another day this place would be a hive of activity. On the highest edge of the property are the house and greenhouses with intensively planted vegetables and hedgerows sloping down to a low point in the field. Though the first few frosts had just zapped the tender crops, the farm was still a mix of summer crops that were hanging on, established fall crops, and new plantings for the winter.

Having just transcribed the interview with the Four Winds Farmers, who started no-tilling over twenty years ago, and then hearing about how Paul and Elizabeth came up with their no-till system on their own a decade before made me think, this is why I'm writing this book. I don't want another grower to have to develop their own system from scratch, when the knowledge and techniques exist. I want to show that there is a way to run a commercial farm that doesn't involve tilling, and I want to make it more likely they will try one of these methods, by showing how others have done it.

Since developing a no-till method through their experimentation, the Kaisers have become some of the most outspoken and articulate proponents of no-till in vegetable production. They have traveled and spoken to growers internationally in addition to holding seminars on their own farm. A quick web search will turn up videos of the Kaisers talking about their system.

Paul and Elizabeth met me by their house, and then showed me around the fields. I asked Elizabeth how they came to be doing no-till.

SINGING FROGS FARM

Elizabeth and Paul Kaiser
Sebastopol, California
Vegetables
*Applied organic mulch (compost),
occultation*

Developing a No-Till Method

"[In the beginning], we weren't even going for no-till. For us, we came at it a very different way, because we didn't have the background as farmers, we hadn't apprenticed on farms, so we didn't have that mindset of how things should work," said Elizabeth.

"Paul and I met in the Peace Corps in West Africa. He was doing agroforestry; I was doing public health, so I was working with women on nutrition, and doing gardening and things like that. After the Peace Corps, we decided to go to Costa Rica where Paul did his masters studies in natural resource management and ecology, and I focused my studies in public health. In fact, I worked as a nurse for the first seven years of starting this farm as a way of making supplementary income as we got going.

"When we started farming here at Singing Frogs Farm, we started with tillage. We hadn't had any direct farming experience. For the people who were here prior to us, farming was a retirement. It was a hobby farm for them but they wanted it to continue on as an organic farm, so they made it possible for us to take it over. They were ex-engineers. He loved his tractors. So he didn't have any perennials because it would impede him turning around his tractors.

"We started with planting hedgerows in the upper fields. I was five months pregnant when we moved onto the property, so our first year we had a single tilled field. The next year we increased to two fields under conventional tillage practices. But the tillage wasn't working, and we hated it. We hated the tractors, we felt like we were killing things. The soil was degrading before our eyes, and we had huge pest problems.

"In our third year we made a change to permanent, hand labor beds. We'd learned John Jeavons' methods in West Africa because that

Elizabeth and Paul Kaiser.

Credit: Singing Frogs Farm

was applicable to teaching somewhere without tractors but with intensive hand labor. We thought, 'We know how to do that.' Also, we had a fantastic employee; he was working on his PhD down at UC Santa Cruz, but working from up here and living on our property starting a chicken business. We thought, 'If we don't have work for Marty over the winter, and we do cover crops like we've done the last two years, we're going to lose him and not have him again in March.' And so we thought, 'Well, it's Sonoma County, we can grow vegetables all year 'round.' So we extricated the tractor, and said, 'Let's make permanent beds, let's do this intensively John Jeavons style."

"You mean with double digging?" I asked.

"We actually didn't double dig. We have two beds that were half double dug by an intern. She and her boyfriend took two months to get two beds halfway done with double digging, then they gave up and just threw compost on the top of the rest, and they planted tomatoes in both of them. Well you know what? There was no difference. We'd done some double digging in West Africa but I don't know anybody that does double digging commercially. We switched to a rototiller. After some time, we started using a little tiller that was more like a Tilther and incorporating a broadfork," said Elizabeth.

"What we noticed was that we were not having earthworms in those beds [that were rototilled]. We had an 'aha' moment. For Deborah Koons Garcia's film, *The Symphony of Soil*, they came to us at the end of filming. They needed all sorts of B-roll: a picture of compost, and a picture of sheep poop because they had some Irish sheep farmer that was talking about her sheep returning nutrients to the soil, and shots of earthworms in a bed where food was actively growing.

"We did an active dig through six beds looking for the perfect veg and earthworm combination, and it was really interesting because it made us say, 'Lots of earthworms. No earthworms. Some earthworms. Lots of earthworms.' And then after they left, we thought, 'So, what's the difference? Oh, that bed that has a lot of earthworms hasn't seen the tiller in a year. That bed with no earthworms was tilled six months ago.' So that was the end of [the rototiller], and then we transitioned

So our transition away from a tractor and then away from a walk-behind rototiller towards a no-till soil management system was just really looking at the biology, looking at the soil's structure, wanting to increase production and keep year-round employment, that led us into the idea that 'tillage is bad.'

—Elizabeth Kaiser

into using a broadfork. So our transition away from a tractor and then away from a walk-behind rototiller towards a no-till soil management system was just really looking at the biology, looking at the soil's structure, wanting to increase production and keep year-round employment, that led us into the idea that 'tillage is bad.'

"There is a video on YouTube by Pesticide Action Network that's about six years old, of Paul in one of these beds here talking about what we're doing, and he doesn't even have the terminology 'no-till.' He's saying, 'We do this non-mechanized, intensive vegetable production in permanent beds.'

Tomatoes and lettuce intercropped later on in the season.

Credit: Singing Frogs Farm

"Soon after that video was released, Paul went to an international conference at UC Davis on climate change and agriculture. What they were saying [about reducing tillage] made sense with what we were experiencing here. The science supported our practices and observations, and then that just sort of led us further down that road, and led us to sharing what we do, and that's what ignited a lot of interest. That has been, in a nutshell, our path.

"There were four things that really pushed us to abandoning tillage. One thing was life. So noticing the lack of earthworms when we were doing tillage. I can remember a specific time when I was holding my baby Lucas, who is now ten, waiting to talk to Paul, and he was using the tractor doing a spring till. And I was standing there waiting for him to finish his pass observing life: the birds over here, and the dog sitting a little ways down from me. He turned off the engine, and it was just this immediate pounce by dog, cat, and a bunch of birds into the field. I realized, 'Oh look, we're tilling up life, and they all love it.'

"Here we are putting up hedgerows and trying to increase ecology because that's our background. We planted our very first hedgerow on the day we moved onto the farm. So ironically, while we were still using the tractor, we were trying to build up life on one side, and then we were killing it on another side with tillage.

"The other aspects that pushed us away from tillage were our climate, the soil quality, the pests, and just the challenges of owning a tractor. Our climate is such that very few people produce in the winter because our rains start in October and go through springtime. In spring you can't have tractors in the field because they'll get stuck in the mud.

Credit: Singing Frogs Farm

Over half of Singing Frogs Farm's compost is created on-site and near their fields from crop residue. When almost mature, it is mixed with purchased municipal compost and composted manure from their county.

"We thought, 'If we want to grow in the winter, we want to feed our community year-round, we have to get this tractor out of here.' We had two tractors and a big truck, and there were several times when tractor A was stuck, tractor B got stuck trying to pull out tractor A, and then, 'Oh bring the truck down in the field and let's hope to not get it stuck also.

"Finally, as I mentioned before, we wanted to keep our employee in place. That was tied into wanting to grow food year-round and have income year-round. The price of living in Sonoma County is pretty high, and our property taxes are about eighteen grand a year."

"Wow. That's a lot! Especially comparing to central Maine," I said.

"So those are the triggers that led us down the road [to no-till]," Elizabeth said. "We stopped using our rototiller nine years ago and finally gave it away last year."

"That's a very good overview of the reasons. I wanted to make sure I understood the actual evolution of your farm. Because no-till wasn't really a thing, when you started, you didn't say, 'Darn it, we're going no-till.' You just stopped tilling and found another way," I said.

"If I started now it would be, 'I want to farm because it's the right thing to do. I want to feed my community, and I want to help save the world, and do the right thing for agriculture and for the land,' and so forth. We didn't understand that aspect eleven years ago. I mean, yes, we understood the aspect of helping build up the ecology, but the whole carbon in the soil issue, we were not aware of at that time, and most people were not," said Elizabeth.

The Economics of No-Till

"Paul and I had long debates on whether we wanted to share what our revenue was on the farm. We finally decided to do it because we felt like people weren't taking us seriously. There are three parts of this farm: one is the ecology, one is the soil and the lack of tillage, and the third is the intensification. And all three of them, to me, are really important to work together. Because if you're not intensive, you're not actively managing that soil all the time, and therefore you get too many weeds, or no green photosynthesizing plants in the ground, and then you lose that soil, and it loses its quality. It's the intensification, doing three to eight crops per year, many of which are multi-species, that feeds the soil and pays the bills. If you can have something super intensive, you don't need so much land," said Elizabeth.

"Our competitors, in the local markets, they need 20 or 25 acres to produce the same amount of food we're producing on 2.5 acres. That's the point; you don't need 25 acres to be a farm. Most organic consultants in the state of California believe you need 20 acres or more to be successful. They don't consider anything under 20 acres a viable farm. Yet we're making as much, or more, food than larger farms. So if you can get yourself a half-acre of land, and bring in eighty thousand in revenue in a year, bingo, go for it. And you don't need to have these huge acreages and mechanization," said Paul.

"Now I also know that there are some people who are much more focused on the economics, and there are some farms that are minimum-tillage, that are bringing in much more than we are. And I've had some of them out here going, 'Why are you growing winter squash?' And we say, well, 'We have a CSA, I know I lose money on it, but I need something heavy in my CSA boxes mid-winter.' And I'm an idealist. I need something to carry me through. Yeah, if we did a lot more cut-and-come-again greens, and mini romaines, and herbs, things like that, we could be even more [profitable], but that's not the farm that I want to manage and live on," said Elizabeth.

"I like it when growers talk about their numbers, it helps people who are considering going into farming. I want to see people be able to make a good living at it, not just scrape by," I said.

"Absolutely," said Elizabeth. "We need to have the conversation, and I need to be able to tell people, 'Yeah, this is what we're making, but please don't focus only on this. If you want to you can make more money, look at [other growers. Others can make] double or triple what we're doing, per acre, and this is where I fit in the spectrum.' I think we need people to see the spectrum."

"I know you said you were making about a hundred thousand dollars an acre—can I say that?" I asked.

"You can quote that. For example, we do outreach as part of the farm also. People ask us, 'You do teaching and consulting and things like that, is that subsidizing things here?' And the answer is, last year we brought in about $335,000, about $32,000 came from the workshops. So it's still definitely a hundred thousand an acre from vegetable production. And we have three acres of production, but that includes the nursery and the greenhouse, and all of that sort of stuff. So if you actually look at the field space, it's two-and-a-half acres, but you need to include farm infrastructure, so that's why we say three acres," said Elizabeth.

"The farm is still paying the bills," I said.

"Of course, an intensive farm this size takes a lot of labor, so most of the revenue

Singing Frogs uses hoophouses over the winter for greens, in March (this is March 15, 2018) they transition to early cucumber and summer squash crops.

Credit: Singing Frogs Farm

goes to people. We have about five full-time equivalents on a farm this size," Elizabeth said.

"Everybody I've visited on these trips is so inspiring. What I want to see is a larger percentage of the food coming from a local healthy source. This is a model that new farmers and people who want to do better, on the triple bottom line, of economic, ecological, and social sustainability, can look to," I said.

Scaling Up

"I was just recently at the Regeneration International General Assembly down in Mexico. There is a lot of talk about the carbon sequestration and so forth. And when I talk about three acres of vegetables, people look at me and they're like, 'Yeah, let's talk about Gabe Brown. He's got five thousand acres. You've got three.' And I say, 'I hear you, and I will absolutely respect the fact that large-scale really positive operations are going to be able to sink more carbon, and that's great. I personally can't have much effect on that, but I also know that we eat vegetables,' said Elizabeth.

"And I also know that we need livelihoods for farmers, and if you look at the food worldwide that is being produced, what we're doing is actually much more similar to what people do outside the United States. It's not a coincidence that we brought in things that we knew from West Africa to farm in Northern California. Seventy percent of the global food is grown by small farmers, on small farms, and it needs to continue that way."

Focus on Your Soil

"In our workshops, we keep coming back to 'focus on your soil.' Depending on who you talk to, you can say there are three, four, five soil principles. To us they are: Disturb your soil as little as possible. Keep your green living plants in the ground as often as possible. Have a diversity of plants. Keep your soil covered. And then the fifth one is to incorporate animals. And we're not directly incorporating animals because we've got these permanent, really high-intensity beds," said Elizabeth.

"So my way of incorporating animals is working with a neighbor who has twenty-four horses that they board using all organic [bedding], and getting the manure. We also occasionally get sheep and chicken manure. I wish there were a better way, but that's how we're doing it. I think you should incorporate them in terms of nutrient cycling. But, anyway, that's what I tell people. It's a little messy here and there. We're not perfect, we're a farm, right?" said Elizabeth.

"Yeah. I have a farm, I understand. There's a reason working farms aren't in *Better Homes & Gardens*. I get it," I said.

"This is true," said Elizabeth.

"I think this is beautiful. Your crops look good," I said.

"There are some weeds back there, there are some flowers that just died with the frost out front, but yeah. It works. We really, really, really try and focus on the production, like trying to get as many of our hours in the production and not in the management. And then we save the management for winter, like putting in hedgerows. We decided to rebuild the hoophouses late because we wanted to spend all our time getting our winter crops in the ground first. And then rebuild the hoophouses once we have time," said Elizabeth.

Leeks and lettuce intercropped.

The Power of Biologically Active Soil

We walked down to the lowest part of the farm, a bowl at the bottom of the property. "These are our seasonal ponds down here. So this is the low point, in that all the water that comes onto the property will eventually end up down in here. And so these have been wonderful for us in learning about and checking our system," said Elizabeth.

"For the prior owner, these were probably just marshy areas. He created [the ponds] by noticing that his erosion silts were coming down here. Before the rains he would take his tractor down here, scoop it up, and put it back on the fields. Which is what he instructed us to do. And we did that our first two years. Then we were like, 'This doesn't make sense. We just want to keep [the soil] up there in the fields in the first place.' As we established the hedgerows and transitioned to permanent no-till beds, that [erosion] just disappeared.

"And then it's been really interesting because in our first few years we'd have one large rain event, say in November, of eight or twelve inches of rain. And boom, these puppies would be full. Now this past winter, it was almost twenty inches of rain before these filled up. That's

Tomatoes and lettuce with hoops for frost blankets.

Credit: Singing Frogs Farm

awesome; that means the rain is staying up there on the fields in the soil.

"And then lastly, we had an article written about us called 'The Drought Fighter' two or three years ago now, by a journalist that really dug in deep. He came and interviewed us and interviewed us, and then he brought an organic consultant out here, and then he took soil samples and sent them to labs. The organic consultants said, 'You know they're using too much compost, they're going to be polluting their water sources.' And Paul and my reaction was, 'Wait, we're using too much compost if you're looking at an organic output system, but not necessarily if you're looking at a biological no-till system. And also a system where maybe you're going to grow one or two crops a year. Not a system where you're going to grow three to eight crops a year [like we do]," said Elizabeth.

"So he said, 'You're going to have runoff, test your water. You're going to have [those nutrients] in there. Test your nitrogen, and your phosphorus down [in the pond water].' And we said, okay. So every year since then, during and after large rain events, we'll test the water running in. We also test the water and runoff at our compost piles, the water at our ponds down there, and then our water here. And what we've found is that phosphorus basically is the same throughout, including our well water. Nitrogen will have a little bit running onto the property, but still safe, within normal drinking quality amounts, and then down here in our seasonal ponds that catch all the runoff for the fields, the nitrates and nitrites will be undetectable. So [we realized], 'No, actually we're not leaching, we're doing the opposite,'" said Elizabeth.

"Your farm is sponging it up," I said.

"Our high soil organic matter is sponging up all the nutrients and we're holding it with all of the soil biology that's there. So, having these ponds has been a really wonderful resource for us in just making sure that our system is working," said Elizabeth.

Bed Preparation and Turnover

I asked Elizabeth about the nuts and bolts of how their system works. She gestured at a nearby bed.

"There's an almost mature Romanesco cauliflower crop. You can see there's some red butterhead lettuce underneath. It's mostly harvested out at this point. So first of all, when we put that crop in, we'll put in two crops, because the cauliflower is a huge plant with spacing two lines, at 24" apart. We want to use that empty space so we multi-crop with a full line of lettuces down the middle [between the rows of cauliflower], and then one lettuce between each [cauliflower] plant. [So you end up with two rows of cauliflower 24" apart, with a row of lettuce between them, and a lettuce plant in the space between each cauliflower plant in the cauliflower rows.] So you're getting about 1.6 crops out of there," said Elizabeth.

"At this point let's assume we've harvested the cauliflower and are ready to transition the bed over to the next crop. When we're done harvesting the cauliflower, we'll chop it off at the ground level or slightly below. I'm going to take some loppers [for very thick-stemmed crops like brassicas]. If I can, I'll use a harvest knife, but it depends on the crop. I'm going to stick them in the ground and chop just underneath the soil surface.

All of that green matter [the bulk of the plant that is left after harvest], then is going to go to the compost, because we're going to retain far more of the nutrients by composting it in a hot compost pile, than by anything else. But we're going to leave all that root mass and rhizosphere in the ground, for obvious reasons: A) not turning up the soil, and B) not getting rid of the most biologically active area.

"And then we're going to assess if we're going to do fertilizers and composts. We don't use a lot of fertilizers. The main fertilizer we'll use is for nitrogen, because we are conscious that you can have too much compost. And we don't want to do that, and it's also a cost for us, for sure. So if you were to put enough compost on to get our nitrogen needs met, [it might be too much]. And yes, we are having three to eight crops in a year so we do need a little bit of extra nitrogen, we've found. We would have far too much potassium and phosphorus [if we tried to meet all the nitrogen needs with just compost]. So we do a pelleted feather meal, and calcium if necessary.

"Then about once every twelve months, maybe eighteen months, we'll do some sort of rock dust, just for micronutrients. Then we'll cover it with compost. When we started out, we were doing a half inch to a full inch of compost on every bed. These days it's as little as it can be, which turns out to be about a quarter to a half inch. We try to just cover the fertilizer so you can't see it. If you have the calcium there it's really easy, because it's bright white. Then we'll transplant in the next crop right away to get our next photosynthesizing plants in the ground. There's no break in feeding the soil, we transition directly from one healthy photosynthesizing crop to the next in minimal time.

"Two questions I often get asked on that whole process: One, don't you have regrowth? And the answer to that is, we really don't if you cut at an angle, under the soil surface. There are a couple crops that will regrow, like radicchio and fennel. And then we'll just manage that, it's no big deal.

"And the second question I get asked is, 'Oh, don't you run into the roots when you're transplanting the subsequent crop? Don't the roots just stay there?' Especially a big, huge root like that cauliflower there, or a Brussels sprout? When we hit [the roots of the previous crop] we'll just move over a little bit, and by the time that next crop is done, and you're into the second succession, [the roots are] gone. I know a lot of people are concerned about it, but they're gone. You can't find them if you want to."

"They're probably also dealing with less biologically active soils, where plant matter doesn't break down as quickly," I said.

"Right, that is definitely true. When we were starting, our soil organic matter was 2.4 percent, and these days depending upon the field, it's 8 to 11 percent. And I think that range is a really great place to be. We have gotten up to 14 percent before, and we let it back off by just planting some very long crops there without adding any compost. What we've found is, at 13–14 percent, the soil was way too fluffy and light, and the plants were just falling over," said Elizabeth.

"I think there's some balance in that area where you've just got really good biology. We haven't done a ton of biology tests, but what we have done shows good predator-to-prey ratios, good fungal-to-bacteria ratios. It's hard to really know what's in there unless you do a genomic test, which is really expensive. Even the testing we have done on our soils, oftentimes scientists are just like, 'Yeah, we just don't know enough about the biology of the soil.'"

"I think that's an area people are still learning a lot about," I said.

"We are learning a lot about it. For me, it's just: feed the soil, have plants in it, green living plants, get

Some examples of no-till intercropping on Singing Frogs Farm: broccoli with direct seeded fava bean in April, and in the background red butter lettuce and brassicas, right tomatoes and lettuce.

Credit: Singing Frogs Farm

the soil organic matter up there to 6–10 percent, and it works itself out. That's my take," said Elizabeth.

Farm Ecology

"I stopped right here because I wanted this hedgerow to remind me to talk about the ecology. Like I said before, I really, really believe in having perennials, and having some natives. If I were to redo our hedgerows, I might put in more fruit trees and other productive plants. The hedgerows are crucial for the beneficial insects they bring in. I had an entomologist out here, and she was looking through the hedgerows and she was so excited. She said 'Do you realize this beetle, he'll go two hundred feet out into your field and feed on pests, and then come back here to home?'" Elizabeth said.

"And then snakes love [the hedgerows. One of our employees] was petrified of snakes when he came here. We have gopher snakes, we have garter snakes, and we have some sliders. They love the hedgerows because they have protection from hawks, they can then come out and hunt, and they can make their burrows under there.

"But he has gotten to the point, he will actually move them if he finds them in an area where they're not needed. There were a lot of beets down at the very bottom of the field, and I chatted with him as those were going in. I said, 'Miguel, we'll have gopher and field mice problems, because

This is April 5th when they started harvesting the same bed as in the photo on p. 281.

Credit: Singing Frogs Farm

they'll chew on the tops of the beets down there. Do you think you could do a little extra trapping down there?'

"Miguel said to me, 'Elizabeth, my friends have been helping me down there. I don't need to set too many traps because of them.' For the last two months, every snake he'd seen on the property he'd put down there. And he was right, we have very few rodent issues in that crop.

"So to us, the hedgerows are not a pretty little thing along the outside; they're an absolutely integral part of the farm. When you have your beneficials, you don't have to spray. Isn't that a good thing? Because then those sprays aren't getting washed down in the soil, and killing your microbiology in the soil. And you don't have to eat them too, right? Win, win, win. So, I so strongly love them.

"We are really super cold down here on this property, but hedgerows offer a huge benefit by stabilizing air masses in our fields. For example, up here, we had five rows of summer squash. One year we had our first fall frost come in, and the two bottom-most beds of summer squash were burned to the ground. The third one was rimmed around the edges, and then [the two next to the hedgerow] had no damage at all and continued to produce.

"So hedgrows just sort of protect the area around them, keeping it a little warmer on those super cold nights and a little cooler on those super hot afternoons. Hedgerows also act as wind breakers, reducing wind stress and moisture loss for our crop plant and the soil. So it's an important part of the intensification and not using sprays, and everything to us. They are a really important part of our system."

Weed Pressure

"Can I ask you about weeding? Because you're building these beds, and you're not going to cultivate them. People are going to want to know how you deal with weeds," I said.

"When we talked about bed transition, one thing I left out is, when we cut out the crop, that is the time when we will pull the weeds," said Elizabeth.

"Some of the things that we will do for weeds—we'll pull them when we take out the crop, and that is our main weeding time. And then when the next crop goes in, we're putting on fertilizer and compost. I didn't say this, but [we only add amendments on] 80 percent of our crop transfers. If it's a short crop, and then another short crop, and everything's great, I'm not going to do [the full bed prep with fertilizer and compost]. But as most of our crops are heavy feeders, and are in there for a long while, we will add some light compost and organic fertilizers.

"When we're doing that, we're putting on a thin layer of compost, just enough to cover the previous soil, maybe half an inch at most, and then most of the time we're putting a transplant in, so this is outcompeting those weed seeds down there tremendously. This process lowers our weed pressure, and as you continue to practice no-till, you're reducing your weed seed bank by not tilling to bring up new seeds," Elizabeth said.

"So is that something that you've noticed as you've gotten away from tilling, is that you're not churning weed seeds up?" I asked.

"This year was so wet, this year was really odd. But in general I would say yes. Our weed pressure has gone down a lot, and we notice that some of our worst weed pressures are in the fields where we did the most tillage in the beginning years," said Elizabeth.

"If something gets too weedy, and we are not able to stay on top of it and get that crop out when it's no longer photosynthesizing at its optimum, i.e., it's older and it's growing more weeds and so forth, then we will do occultation. We do occultation with landscape fabric. I know most people use silage tarps, but we trialed silage tarp and don't see much of a difference. Landscape fabric works really well for us and they're more breathable," said Elizabeth.

As we walked back up toward the house, Paul rejoined us. "They're actually permeable to both moisture and gas exchanges, they're lightweight and portable, and they last a really long time," said Paul.

"So in some of our lower fields there were a couple beds down there that were covered. That was because we'd let those go too far because of labor issues that we were having. It has been a hard year labor-wise for us. In the winter we do far more covering, because often we're not able to get another crop in the ground during November or December due to limited daylight for photosynthesis. Therefore we cover the beds to keep the soil healthy and protected from the elements," said Elizabeth.

"To be able to say that you just weed when you're taking the crop out, that's a huge improvement right there," I said.

"It's not even weeding, it's just transitioning the bed," said Elizabeth.

"It's just clearing the bed, which means you pull out a few stray weeds here and there, and you cut the crop out, done," said Paul.

There is a time-lapse video of the bed turnover process on Singing Frogs Farm, which is well worth seeking out online.

Referring to the video Paul says, "We said in the video, it's a 45 minute time lapse, and it's three people, four 85 foot long beds. So for 340 feet of bed, it took three people 45 minutes to go from the standing crop they harvested that morning, to transplanting the next crop. So very fast, easy, and such a smooth process with no disturbance of soil, and no loss of food going into the soil, referring to the root exudates that come from a living plant and feed the soil life. You have photosynthesis [from the old crop] in the morning, and you have photosynthesis [from the new crop] in the afternoon, and this constant ongoing process with

Credit: Singing Frogs Farm

Elizabeth giving a farm tour.

no disturbance and no break in the health of the soil and continuity. That's the idea," said Paul.

"My thought is, even if people don't care about all the soil life and sequestering carbon and all that stuff, no-till methods have value because they're a simplification of the farming process. I don't know if you've read Ben Hartman's book, *The Lean Farm*, but he's always talking about eliminating extra work. He calls it 'muda,' it's a Japanese term for any kind of drag, any inefficiencies, and so I'm thinking about no-till and I'm reading this book and I'm thinking, tillage is just muda," I said.

"Weeding is muda," said Elizabeth.

"It's anything that's not directly contributing to making a profit, to the success of the farm. I think the problem is that tillage is a paradigm. Right? Most people can't imagine farming without the tillage aspect, and that's what I want this book to do is say, 'Here are the people doing it, here are the examples. It can be done. And if you can get rid of it, it will streamline things to do so,'" I said.

"Absolutely. I said it in different words when we were down there, but I said we really try and keep all our labor on production and not on maintenance," said Elizabeth. "And then some of the larger projects, like rebuilding the hoophouse over there, that's going to happen in the winter. To keep people employed, because some of those things do need to happen."

"But by late January, they're already planting and harvesting, and planting and harvesting all week long, and so by late January our winter's over. We have to have winter projects done by mid-January otherwise we'll never get them done. We're already up and running in terms of production. We never stop production. We hit the ground running hard in late January and February, when other farmers in the neighbor-

The radish harvest—Singing Frogs does more direct-seeded crops in the spring including radishes, mustards, and salad turnips. Over winter they direct seed lettuce mixes and spinach. Throughout the year they direct seed carrots.

Credit: Singing Frogs Farm

hood are still waiting two more months to even consider doing tillage," said Paul.

"One thing we always point out is, a while ago I was on the board of the Santa Rosa Farmers Market. It's the biggest market in the county, it's year-round, and in mid-May, there are on average fourteen or fifteen farms selling produce. By June 15 there are forty-eight. So the number of farms from mid-May to mid-June triples, that's when all of the farms that are tillage-based come online. What happened to January through June for eating food?" said Paul.

"In January I counted, there were four. And two of those, half of their booths were sprouts," said Elizabeth.

"And it's even more of an advantage to not have to do that tillage in a place like Maine. We have a wet winter, we get a lot of snow. And then it melts and there've been so many times where we have put off our planting plans because we couldn't get on the field to till. We have a very short season, so if you miss two weeks at the beginning, you've missed a lot of the most important part of your season," I said.

"But if you can whip a blanket off the bed, when there's still a little bit of snow on the blanket, and get in there on your feet without a machine, put down some compost, transplant, you got crops there. Put a frost blanket back over them. And harvest in a few weeks. Much nicer," said Paul.

"Yes. Absolutely. So we find, a huge profit center for us is March, April, and May," said Elizabeth.

"Plus, I imagine if you're there at the beginning of the season, there are probably a lot of people who are going to like your stuff at the beginning and stick with you for the rest of the season," I said.

"Exactly. Keeping year-round CSA and year-round market space, you don't have to re-find [your customers] all the time," said Paul.

"Yeah, we do a weekly CSA from May through Thanksgiving, and then we do every other week over the winter. Because we do have a little bit less coming out of here. People also like their winter veggies a little bit less than they like their summer and autumn veggies. Which works as a win-win, because then we can take more to market, and our customers, they're used to coming to us. And restaurants. You want to get into a market that's challenging to get into? You show up to the market manager in February, 'Hey, guess what I got?' And boom, you're in. Or a restaurant. 'Hey, I've got…blank blank blank,' you're in, and then they keep you. So it's a huge marketing tactic," said Elizabeth.

"Yeah, that makes a lot of sense. Your no-till method is another way to extend the beginning of the season, almost like growing in a greenhouse, but without all the infrastructure," I said. "I just wanted to ask, do you have any advice for aspiring no-tillers, people who want to get started with this system?" I asked.

"Absolutely—start small. And don't let things get out of control," said Elizabeth.

"You don't want to start ten acres of this. You'll grow far more vegetables off a half-acre then you will on ten acres if you're intensive and do it well," said Paul. "So start with a half-acre. Start with one acre. Because it's really one and a half people per acre to manage and produce. And go from there, that gives you a lot more free acreage for chickens, ducks, cattle, sheep, whatever you want. But really focus your vegetables on being intensive and do it well, so it isn't a failure and so that all the beds are constantly being produced from. Because that actually feeds the soil by being constantly productive, and then as you get really good at it, keep adding more beds or more fields and expanding outwards and more crew members maybe."

"Another thing I would say is that the nursery is really super important. And having really resilient, large, healthy transplants that go in the ground is of key importance. We also really stress with new farmers: figure out how to get really good healthy transplants out there. That is another thing that is super important to us," said Elizabeth.

Before I even arrived at Spring Forth Farm in Hurdle Mills, North Carolina, I got an idea why they might need to build some soil. On the way in to visit on a warm day in early May, I passed a tobacco transplanting crew. The farm was surrounded by acres of freshly planted tobacco, which from a distance looks like cabbage seedlings; little green flags waving in a sea of red.

I immediately recognized the brick-red clay soil from where I grew up in Virginia. Jonathan and Megan Leiss run the farm, and they were kind enough to accommodate me in one of the busiest weeks of their year—on the Monday before Mother's Day. I arrived at an exciting time. Megan and Jonathan were putting the finishing touches on the beautiful passive solar house they had built themselves and recently moved into. For the previous three years they had been living in a small trailer on the property while they built the house.

Megan was two weeks away from quitting her teaching job in order to farm full-time. The Leisses grow flowers which they sell wholesale to local florists, operate a flower CSA, and grow food for themselves and a local food pantry mixed in with the flowers.

Indeed Megan confirmed the soil on their farm was pretty much the way it looked on the drive in—more like clay than soil, with an organic matter content of 2 percent when they bought it.

"This land has been cultivated in tobacco for over a hundred years. And when we bought this land it was in tobacco. So it has just been run into the ground. If we didn't have some sort of mulch cover, we would till a bed and then within an hour on an August day you could walk on it like it was concrete. It became so hard, instantly. So we had to figure out something else that would help us at our scale and size to build and take care of our soil. We decided that no-till farming was the way to quickly build organic matter and soil biology," said Megan.

Jonathan and Megan Leiss
Hurdle Mills, North Carolina
Cut flowers
Occultation with silage tarps and permanent beds

They use silage tarps to smother weeds and break down biomass left over from previous crops and cover crops. To show me what they're dealing with, Megan pulled back a tarp, scraped away the decaying cover crop residue on top of the soil and dug out a handful of soil. Her fingers left smear marks in the earth. She handed me the lump of clay, and I squeezed it in my hand. You could have made a pot out of it.

The previous day, Megan was transplanting. Finding a worm was a joyous occasion. The good news is that clay soils hold a lot of nutrients, if they can be enriched with enough organic matter.

The extremely low amount of organic matter complicates almost every task, making it difficult to draw soil up around the roots when transplanting. Percolation is very slow, so rain and irrigation tend to pool and run off instead of soaking in.

No-till is how the Leisses are dealing with these soil challenges without the use of a tractor or other heavy equipment. When I visited, they were cultivating one acre without any type of motorized farm equipment. Since then, they have purchased a BCS two-wheel tractor from their neighbors. They use it to flail mow cover crops, which helps them break down faster during occultation. They also got a power harrow for the BCS, which they plan on using on a very shallow setting ("less than 1") to help get their cover crops to germinate.

Unlike rototillers that have tines mounted on a horizontal axis, mixing soil layers, a power harrow has tines mounted on a vertical axis, mixing soil without inverting the layers. Power harrows are also less likely to create a plow pan, since there are no tines dragging horizontally through the soil.

"The power harrow does not invert the soil, it 'stirs' the soil like an egg beater leaving the soil structure intact," said Jonathan.

Megan and Jonathan with some of their flowers.

Credit: Spring Forth Farm

When they really need a tractor, neighbors collectively have eight tractors that can be hired to push around a particularly large pile of organic matter or help with any other job that goes beyond the human scale.

Looking at their land with them, two advantages of no-till came to mind. First, a tractor would be too big. Their one-acre farm doesn't justify or demand that much horsepower. So they skip the tillage and prepare the soil for planting by hand. This is particularly useful in wet weather. "When other farms are not able to work the soil yet because it is too wet, we can prep beds, no matter the weather," said Megan.

"Farmers who cultivated our land previously spread around some of our most noxious perennial weeds with their heavy equipment. We are trying to break that cycle and stop spreading weeds," said Jonathan.

Another advantage of going tractorless is that traditional tillage might well make their soil worse. While a plow or rototiller could serve to mix some organic matter into their clay, it would also beat the soil structure up further, potentially making the soil even tighter and creating a plow pan at the bottom of the tillage zone.

The year I visited, 2017, was the fourth season of production for Spring Forth Farm. They used minimal tillage at first and then moved to 100 percent no-till farming in 2015. Though soil organic matter has increased over three years, lingering challenges include the fact that the soil is still low in organic matter, and certain weeds persist through the occultation process, most notably nutsedge and dock.

Credit: Spring Forth Farm

A cover cropped field before occultation (*above*) and the same field after occultation (*below*).

Making Beds

"When we started, we hired our neighbors with their tractors to break new ground for us and build our beds with their bed former. We currently maintain them as permanent raised beds, adding organic matter through cover cropping and composting," said Jonathan.

"If we could start all over again, we would have dumped huge amounts of organic matter in each section, disked it in and then laid out beds," said Megan. "We did this in our 96' × 30' hoophouse we just built. We added four-and-a-half yards of leaf mold and one yard of compost per bed."

"One thing that we had hoped for that hasn't worked out the way we wanted is the seed bank is not being depleted under the tarps," said Jonathan. "The fescue seed is being depleted, but that was our fault because we planted fescue and let it go to seed while we were building the house! But a lot of the weeds that we have trouble with, the tarps don't cause them to germinate [and die]. The biggest ones are smartweed, chickweed, purple henbit, bindweed, and a little bit of crabgrass."

(When I checked back in with them almost a year later, Jonathan said, "This has actually gotten much better since last year. Perhaps occultation just takes time to burn through the seed bank?")

In the meantime, they are putting landscape fabric down over the beds and planting through holes into it, to suppress weeds during the growing season.

"One thing that has been really interesting for us is how many different no-till systems there are. We've been trying to pull from as many different ones as we can," Jonathan said. "After hearing the interview with Patrice

Beds with landscape fabric are ready for planting after occultation on the left, and beds still under occultation on the right. In the middle is part of a bed with the tarp peeled back to show what it looks like under the tarp.

Credit: Spring Forth Farm

Gros on the Farmer to Farmer Podcast we started using a lot of straw."

Jonathan gestures to the neighboring land surrounding the farm. "All of this over here is wheat, in a tobacco/wheat rotation. So wheat straw is a good local resource and they will deliver it for $2.50 a bale and stack it. We can have as much wheat straw as we want, but it has its own problems. Now we've been having issues with slugs because of it. You know, if it's not one thing it's another."

"However, since we stopped tilling, our soil life is exploding, including a healthy population of ground beetles and signs of earthworms. If we break open a clod there will be tunnels from earthworms, which we've never seen before," said Megan. "The ground beetles that we are now finding in the soil eat slug larva. This is a great example of how no-till farming helps us to return to a more natural and biological way of farming. Our aim is to work with nature, not against her. Nature already has perfect systems in place to create healthy environments for plants. We want to mimic these systems so that we can manage our farm in a more natural way."

"So now we have a slug population, and a ground beetle population. We're starting to get all these other insects and wildlife. When we first planted in March we lost some plantings to slugs almost completely. Now the ground beetle population is eating more of the slugs. They're catching up and we're not really having an issue now," explained Megan.

Credit: Spring Forth Farm

Mulching with wheat straw.

Bed Prep

Megan was in the middle of preparing beds for late spring planting so she showed me what she had been doing. "In the spring, the tarps are removed, the beds are forked, we add a quarter to a half-inch inch of compost and fertilizer, drip lines and landscape fabric are put down in

Credit: Spring Forth Farm

Jonathan in a block of peonies.

preparation for planting. We use Harmony fertilizer and also regularly inject fish emulsion through our drip lines throughout the season. We add minerals and micronutrients based on recommendations from our soil tests."

After pulling back the tarp that had been on over the winter, she forked the beds to loosen the soil. I was surprised when she showed me that she had been using a garden/potato fork instead of a broadfork.

Megan explains, "We had been using a broadfork, but I found that it was just so heavy and cumbersome that I just like a little potato fork better."

"The other thing about the potato fork is that with the broadfork, you have to step on the bed, because you're walking backwards down the center of the bed. But with the potato fork you can do it from the sides," said Jonathan.

"I started trying out the garden fork because it was a lot lighter for me to use and I felt that I could actually get it deeper into the soil than the broadfork, even though the tines aren't as big. I'm also spacing the forking closer together because it goes so much faster. I think it's twelve here a dozen there. It's getting aerated either way," said Megan.

"I uncovered these three beds today. I was able to fork all three, with just a little potato-digging fork in an hour. Also, I don't have to bend at all. I'm just shoving it in and cracking it, shoving it in and cracking it. I go down one side and back and I can do three 30" × 80' beds in an hour," said Megan. "So that's a really quick thing we can do in any soil conditions. It just dumped rain last weekend and if we were working with a tractor we definitely couldn't be getting into our beds at all. But now we can fork, fertilize, put down Sluggo, put down landscape fabric and drip tape. Then it's ready to go and it can sit here for a couple of weeks before we actually plant it if we need to."

"At our size we do a lot of half-bed plantings, and plant the other half later to something else. It's nice, I know a lot of farmers prep the beds in the fall and then plant in the spring. And this is one of the ways we can try to get ahead, to uncover a bunch of beds at once and just fork them before planting."

Bed Turnover

At the end of a crop or season, they pull the irrigation lines off the beds and flail mow the remains of the crop. "We take up the irrigation lines because our mower just tears them up," said Megan. "We then immediately pull the lines back onto the bed and either put out occultation tarps or seed cover crops. If we are cover cropping, we mow everything down, portion out in buckets per bed the cover crop seed we're planting. Sprinkle it on the beds, add one bale of straw per bed, and let it rain."

"We recommend one bale per bed because two bales per bed is really too thick. With two bales, some of the transplants don't even come up above the level of the straw. It's a really great environment for the slugs. They like that," said Megan.

"So we put the seed on the surface like planting grass, then put straw on top of it, and we don't work it into the soil at all. We've had good luck with that with tillage radishes, oats, and clover, but it doesn't work so well for getting buckwheat and other summer cover crops to germinate," said Jonathan.

One thing about no-till is that it takes longer to make changes to the soil. "This is one of the challenges," said Jonathan. "Some amendments you can just spread on top of the bed or rake it in and they're fine. Some stuff isn't going to work the way you want it to without having a more mechanical way of working it in."

"The straw is our solution for getting cover crops to germinate, because we can't till the seed in, and we have just invested in a power

Credit: Spring Forth Farm

Jonathan showing how some perennial weeds like this dock with a lot of energy stored continue trying to grow under the tarp.

harrow for our BCS so we are hoping this will help our summer cover crops establish better. What we've currently been doing is, we broadcast the seed on the surface and then spread one bale of straw."

"And of course the wheat straw we get is also germinating. So it's adding some cover crop," Megan adds.

"That's the system that works pretty well for establishing winter cover crops," said Jonathan.

Flowers and the Philosophy of No-Till

Sunflowers and other crops planted through landscape fabric.

Credit: Spring Forth Farm

"There's been a lot of interest in high-dollar, high-intensity vegetable production systems which rely on baby salad mix," said Jonathan. "Well, a lot of these systems use the same basic principle which is a short turnaround crop. If you can get it in and out in 20 days and plant something else the same day, that's the basis for making money in that system. This doesn't work with flowers. You could turn flowers around quickly, but there's no 20-day flower crop," said Megan.

"Sunflowers are as close as you're going to get. They're 45 or 50 days for us. We haven't even tried to double crop one right after another because we try to get a cover crop in there between flower crops.

"No-till makes it possible to start a farm with virtually no debt. We are very much no-debt people. What took us so long to build our house is that we built it debt free. And we built the cooler and the germinating room out of the profits from last year. We're just reinvesting everything back into the farm.

"But if we had to go out and get a tractor, even one that wasn't expensive it wouldn't be possible to run debt-free. We are also able to eliminate the time and money we would spend maintaining and fixing a tractor. Investing in a tractor would make us have to scale the business up. No-till farming allows you to have a viable small-scale business without a lot of payments. One of the nice things about no-till is that you can try it on for size, without the commitment of buying a tractor or even a two-wheel tractor. It is a perfect way to get into farming if you aren't certain you want to make it a career.

"We don't have any intention, at least right now, of hiring people. We want to keep the farm the size it is so Jonathan can keep working at the fire department with me here full-time.

"Our systems we have set in place make it possible to return to that family size, family-run farm. I've worked for a bunch of different farmers, on much larger operations than ours; they needed six or seven people to effectively run their farms. Labor was unreliable, people would quit in the middle of the season leaving the farms shorthanded. After working for other farmers I realized that I was not interested in

One of Spring Forth's flower fields.

Credit: Spring Forth Farm

Credit: Spring Forth Farm

One of Spring Forth Farm's CSA bouquets.

that size farm for myself. No-till farming was the key for us [to be able to operate on this scale]," said Megan.

"So we're not making gobs and gobs of money but we also don't have a lot of overhead and we are choosing to reinvest our profits in our business right now."

When I ask where they got their ideas for no-till systems from, Megan says, "We've been pulling from a lot of different systems. We tend to use Jean-Martin Fortier's cover crop ratio recommendations. We got the idea about using straw for cover cropping from Patrice Gros. We're using occultation to terminate cash crops and cover crops by smothering like Tony and Denise [Gaetz, of Bare Mountain Farm in Oregon (see interview p. 123)]. We have experienced a lot of trial and error to come up with the systems we use. We are always looking for new information and resources that can be transferred to our farm."

"After reading the Neversink Farm article ["Welcome to the No-Till Revolution" in the March 2017 *Growing for Market* magazine, page 260 of this book] we're really interested in getting more topsoil in and figuring out a way to effectively get more organic matter on the tops of the beds. I really think what we're going to end up developing for Spring Forth Farm, is a hybrid of all of these systems," said Megan.

I MET BRYAN O'HARA ON HIS TOBACCO ROAD FARM in Lebanon, Connecticut, on a hot day in June. Little did I know it but in two days I would see his system more or less replicated on Natick Community Farm. The audio of this interview was not recorded, so there is less of Bryan's voice than in some of the other interviews, but the gist of the visit was preserved through my notes.

Bryan was kind enough to show me around at a very busy time of the year. According to the calendar it was still spring on June 12, 2017. But according to the thermometer it was summer. It was unseasonably hot and I'm sure Bryan had a lot of other things to do besides show me around. Nonetheless he took me out to a field by his house in an unhurried and friendly manner.

Tobacco Road Farm grows vegetables on about three acres, with half in year-round production, and the other half cover cropped over the winter. Bryan pointed out that he has made his living from the farm for more than twenty years, and that the field we were looking at has been cropped year-round for that entire time.

Bryan has developed a farming style that involves no-till methods, on-farm composting, and preparations from Korean Natural Farming. He has carefully refined his methods over the years. However, for the purposes of this book we are going to ignore most of the other interesting stuff and focus on his no-till system.

Bryan's basic system involves solarizing a previous vegetable or cover crop, applying compost (and maybe Korean Natural Farming farming preparations at the same time—go look those up if you're interested), direct seeding or transplanting into the compost, and covering the compost with mulch to protect the biota from getting fried in the sun.

"Now, the earthworms eat a lot of the mulch," says Bryan, echoing a refrain I've heard from other no-tillers. As soil life has proliferated,

TOBACCO ROAD FARM

Bryan O'Hara
Lebanon, Connecticut
Mixed vegetables
Solarization and compost mulch

Rows of peppers and other crops planted through mulch inside and out of a hoophouse on Tobacco Road Farm.

Credit: Andrew Mefferd

there are more mouths consuming the organic matter going onto the soil.

"We use more mulch (made of chopped straw, leaves, and wood chips) and less compost than previously. Under tillage, compost usage was probably 60–80 tons/acre per year. The tonnage of mulch now used is probably about equal to the compost—roughly 30 tons/acre of each. It's very useful to feed the field biology," said Bryan.

For better or worse, some of those mouths are slugs. One of the challenges to methods that leave more organic matter residue on the surface is that they tend to generate more slugs, which come for the decomposing organic matter but stay for the crops.

One thing I've heard from no-tillers is that the populations of ground beetles and other slug predators tend to catch up with the slug population as the soil is disturbed less. In the meantime, slugs may take a heavy toll on crops. One short-term solution is Sluggo, an OMRI-listed slug and snail bait/killer whose active ingredient is iron phosphate.

Bryan suggests managing slugs by leaving mulches off during particularly moist parts of the year (when they would be less important anyway), using solarization to kill the slugs between crops, and avoiding the lush growth slugs and other pests are attracted to from having too much nitrogen.

Bryan developed his no-till system because over time, many things on his farm were getting worse. Yields were going down and his soil was holding less water and drying out quickly in the summertime. Tillage was reducing fungal activity, and along with the mechanical action of crushing soil structure this was contributing to poor soil aggregation. Weeds were becoming more numerous and pest and disease pressure was increasing. He tied all these effects back to the worsening soil.

His current no-till method is the result of lots of experimentation over many years. His aim throughout has been to build the soil up instead of burning up the biology and organic matter through tillage.

"Tillage is a nutrient flush from all the death you just wrought on the soil. Over time, the soil runs out of steam," Bryan explained. You are sacrificing some of the organic matter and life in the soil to the growth of the year's crop. Without replenishing it, you eventually run out of organic matter, soil biota, and fertility. This can be seen when diligent cover cropping and compost use fails to increase organic matter percentages significantly year-over-year, because tillage is burning up as much or more organic matter than is being added to the soil. "Tillage doesn't give you nutrient balance, it gives you nutrient release," said Bryan.

He developed this system by gradually reducing tillage over time, transitioning entirely to no-till in 2012. The first step was to stop deep tillage, and then to implement a permanent bed system with shallow tillage. Next, Bryan went from a rototiller to a field cultivator. Finally, he just stopped tilling altogether. "Now, no tractor goes in this field," he says.

A row of transplanted celeriac.

Organic Mulches

One of the foundations of Bryan's system is his high-carbon compost, which he makes on the farm and applies at a rate of roughly 30 tons/acre/year. The main ingredients are cow manure, vegetable scraps, wood chips, leaves, and straw. He also adds silica and calcium as needed, and a wide variety of other materials to make custom blends to meet the fertility needs of specific crops.

"Our fields are nitrogen-rich from biological fixation and previous organic matter application. Carbon helps balance this. It also helps feed fungus, and a high percentage of woodchips, roughly 40 percent, greatly aid in passive aeration of compost piles," said Bryan.

Another important part of the system is the straw, leaf, and wood-chip mulch Bryan uses to put down over the compost, to maximize the amount

Credit: Andrew Mefferd

of microorganisms that survive in the compost after application. He tries to apply the mulch quickly after applying the compost to keep it from drying out.

Bryan uses a bale shredder to chop straw into smaller pieces to make it easier to apply to the beds. To deal with weed seeds or leftover grain in the straw, he lets it get rained on and allows the seeds to sprout after chopping. When the chopped straw is gathered later with pitchforks, the sprouted seeds die.

Solarization on Tobacco Road Farm. This can work in just 24 hours under warm conditions.

Credit: Andrew Mefferd

The benefits of sticking with a system like this are visible when Bryan shows me a bed that has been planted four times over the last three months, and hardly a weed has germinated. Using transplants do speed bed turnover, but his quickest crops are usually direct-seeded salad greens.

"Most of the home fields are in year-round vegetables and some of it grows four or more crops a year. A rotation that fast is not common but certainly does happen," said Bryan. "Four times would require three harvests in three months, so that would probably be a salad green like arugula, to pea shoots, to lettuce, to a fall crop like spinach."

We look at a bed of cilantro that has never been weeded. There's an occasional weed, but nothing that would cause the cilantro any problems. The minimal weed pressure in this unweeded bed would be excellent for a bed that had been weeded mechanically. To have it in a bed that has never been weeded is almost unheard of.

Direct Seeding

Seeding the cilantro we were looking at was simplified by the fact that it was planted directly after another crop. Being able to go from one crop to another without having to go through the whole process of tilling and remaking beds is a huge advantage of no-till. Because weed pressure is so low, and because the soil remains light, the cilantro could be planted as soon as the previous crop was removed.

For example, the cilantro we were looking at was planted with an Earthway seeder, though any other single-row seeder would do. After removing the previous crop, Bryan used a hoe to create six furrows in the top of the bed. Then he ran the seeder in the furrows and covered with a thin layer of compost and straw.

Bryan often sows by hand. For scatter seeding in beds that were cover cropped, he mows the cover crop down using a rotary mower on a walk-behind tractor. Then the beds are solarized to kill the cover crop, after which he applies a thin layer of compost (about a half inch, little enough so that you can still see soil through the compost) over the top of the bed to both fertilize and provide a loose surface to plant into. Finally, he scatters seeds onto the top of the bed by hand.

To get enough soil contact for the seeds to germinate, Bryan designed his own implement to cover them. I know from experience that seeds sown on top of the soil and left to sprout don't typically germinate very well. To solve the problem of how to direct seed into a bed without tilling or drilling it in, Bryan made a hand-held ring drag that duplicates the soil-covering action of the rings on the back of seed drills. To do this he attached the rings for a seed drill (available as "grain drill drag chains" inexpensively from Agri Supply, among other places) to a long tool handle. The resulting tool is like a rake with drag chains instead of tines.

To cover hand-scattered seed that has just been planted in compost, Bryan pulls the ring drag down the length of a bed and back. That is usually enough to settle small seeds into the soil to germinate. Another option for covering larger seeds is to use a Garden Weasel, a hand-held cultivator with rotating tines, to churn seeds into the soil, and then run the ring drag over as well.

Sometimes furrows are ripped through the high-residue surface, then a seeder is run down the furrow, or seed is placed by hand and soil pulled back over the furrow. Yet another option for covering seeds with compost is to straighten out the tines of a leaf rake, or use a tool like the Groundskeeper 2 rake instead of the ring drag. But Bryan has had the best results with the ring drag.

Then he covers the compost and seeds with a thin layer of chopped straw, leaf, and wood-chip mulch—not enough to smother the seedlings when they germinate, but to mostly shade the compost. Bryan points out that the advantage to hand-sowing some seeds is that planting densities on crops like carrots and scallions that don't have a big canopy can be much higher than when planted as usual in rows.

These crops can be planted densely over the entire bed, not just in the rows, if bed space doesn't need to be sacrificed for unplanted areas for cultivators to go (i.e., the space between rows). This gives Bryan's hand-seeded beds a much different look from most vegetable farms. We walked by many beds with tiny seedlings growing evenly across the top of beds—not in straight rows—from below a layer of mulch. Transplanting follows more or less the same procedure, by planting into a layer of compost, followed by mulching.

A maturing bed of direct seeded red lettuce.

Credit: Andrew Mefferd

After years of layering the beds, there are so few weeds that there is usually little weeding necessary between planting and harvest. As in other no-till systems, this allows growers to focus on production (planting and harvesting) rather than waste time on tilling and weeding.

"With no weeds, cover crops can be broadcast into vegetable crops at any time, which is very versatile, and makes cover crops much easier to work with. For example, crimson clover in August at 50 pounds/acre, followed by winter rye in September at 250 pounds/acre right over fall crops. When the fall crops are harvested they leave behind a beautiful cover crop," said Bryan.

He times his plantings so there is almost constant vegetative cover and maximization of yield—as some things grow in, others are harvested out.

For example, Bryan's young plantings of cucumbers and melons are bordered by spring peas. Once the peas are harvested, they will be mowed, solarized, and mulched by the time the cucurbits vine into the pea beds, getting a pea crop off of the area that otherwise would just be sprawl space for the vines.

Bryan tells me that he has been able to exceed the yields from traditional cultivation on every crop except potatoes, but he's still working on it. There are other benefits that he has noticed since going no-till. He used to have 5-percent stem rot in garlic over the winter; now, he has hardly any. The water-holding capacity of his soil has been greatly increased, to the point where he got rid of his drip irrigation system.

"Irrigation labor and equipment savings have been significant the last few dry years, and extremely productive, with sun all the time, and the roots hydrated by well-structured soil and mulch," said Bryan.

Solarization

Another foundation for Bryan's system is solarization to quickly terminate cover crops without tillage or herbicides. "A big breakthrough was quick 24-hour turnaround solarization," said Bryan. He can do solarization this quickly roughly from April to October in his climate, on sunny days when it's at least 70 degrees. In borderline conditions—when the weather isn't warm enough or the day isn't completely sunny—he has to leave the plastic on for longer.

This is where readers will have to adapt practices to their own conditions. In times of the year that are not warm enough for solarization, vegetation may be hoed or simply mulched over with a layer of compost to smother it. Though solarization does a good job of killing mown cover crops and annual weeds, it will not kill deeply rooted perennial weeds; these have to be pulled.

Bryan wondered about the effect of solarization on the life in the soil, since the whole idea is to disturb it as little as possible. So he poked a thermometer probe through the plastic during solarization to see what the soil temperature was. He found a 50-degree difference between the air temperature and the temperature under the plastic (from 75 in the air to 125 under the plastic) and about a ten-degree temperature gain at one-inch soil depth, and little temperature increase below that. The soil temperature was very high at the surface but quickly cooled off as he pushed the probe in deeper, leading him to believe that the effect is mostly confined to the cover crop on top of the soil, including the slugs.

Bryan has found that Tobacco Road Farm is more profitable under the no-till system. The harvest efficiency is much higher, with more bounteous crops and almost no weeds. And if anyone should doubt the level of productivity, profitability is excellent—along the lines of other no-till growers I have talked to.

In addition to improving the bottom line, there is greater diversity of life visible in the soil, including more earthworms and more fungal mycelium. Water holding capacity has increased, and there is better drainage and quicker infiltration. Erosion has been decreased because the soil is almost always covered, and never churned up.

For growers wishing to get started with no-till, Bryan advises to "experiment with the system so you know the ropes, so when it's successful you can launch into it." If they have been using tillage, he suggests using it one last time to try and clear as many of the perennial weeds as possible, since these will likely not be killed by 24-hour solarization.

I hope that Bryan derives a great deal of satisfaction from the fact that there are others out there doing no-till because of his work to promote it, such as Casey at Natick Community Farm, who cites Bryan as the impetus behind his methods. And as I noted in the beginning, seeing Bryan speak at a NOFA-VT conference was one of the events that showed me there was a critical mass of these farmer-developed no-till methods.

There is a great deal of complexity that has been refined over many years in the farming system at Tobacco Road Farm. Explaining his no-till system is just scratching the surface of what Bryan is doing. I was very happy to hear that he is working on a book about his farming methods, including all the good stuff we didn't have space for here. Keep an eye out for it in the near future.

Notes

1. climate.nasa.gov/causes/
2. epa.gov/ghgemissions/global-greenhouse-gas-emissions-data
3. e360.yale.edu/features/soil_as_carbon_storehouse_new_weapon_in_climate_fight
4. ipcc.ch/pdf/assessment-report/ar5/wg2/drafts/fd/WGIIAR5-Chap7_FGDall.pdf
5. un.org/sustainabledevelopment/blog/2017/06/world-population-projected-to-reach-9-8-billion-in-2050-and-11-2-billion-in-2100-says-un/
6. rodalesorganiclife.com/garden/no-till-gardening
7. soilquality.org.au/factsheets/organic-carbon
8. extension.umn.edu/agriculture/soils/soil-properties/soil-management-series/organic-matter-management/
9. hobbyfarms.com/how-no-till-farming-will-help-you-save-water/
10. David R. Montgomery and Anne Biklé, *The Hidden Half of Nature*, p. 2.
11. David R. Montgomery, *Growing a Revolution*, pp. 26–28.
12. David R. Montgomery, *Dirt: The Erosion of Civilizations*, pp. 222–24.
13. USDA-NRCS, "Farming in the 21st Century: A Practical Approach to Improve Soil Health."
14. forbes.com/sites/bethhoffman/2013/07/02/gmo-crops-mean-more-herbicide-not-less/#50a1532f3cd5
15. Jean-Martin Fortier, *The Market Gardener*, pp. 104–06.

Resources: No-Till Tools and Supplies

Mulches

- Reuse is probably the most economical as well as the least consumptive manner of acquiring tarps. A nonprofit called Cleanfarms Inc. brokers used silage tarps, the black on one side/white on the other heavy plastic covers used by many no-tillers for occultation. For more info, contact info@cleanfarms.ca.
- Large sheets of cardboard are often available for free from furniture, bike, or other stores that receive large packages. Remember that only cardboard with black ink is currently acceptable for certified organic growing.
- The cheapest way to get clear plastic for solarizing is used greenhouse plastic. If you know someone who is going to be recovering a greenhouse, ask for their plastic!
- Farmer's Friend has pre-punched ground cover and fabric silage tarps. farmersfriendllc.com
- FarmTek has a wide selection of tarp materials and related supplies. farmtek.com

Drop Spreaders

These companies make drop spreaders that drop compost or other loose mulches more neatly than a traditional manure spreader.

- Earth & Turf Products earthandturf.com
- ABI Attachments abiattachments.com
- Millcreek Manufacturing millcreekspreaders.com

Roller-crimpers

- The Rodale Institute maintains a list of tractor-mounted roller-crimper dealers, plus building plans if you want to build it yourself. rodaleinstitute.org/why-organic/organic-farming-practices/organic-no-till/
- Earth Tools sells a roller-crimper that mounts on a walk-behind tractor. earthtoolsbcs.com

Books and Magazines

- *Organic No-Till Farming* by Jeff Moyer
- *Dirt* and *Growing a Revolution* by David R. Montgomery
- *The Hidden Half of Nature* by David R. Montgomery and Anne Biklé
- *The Earth Moved: On the Remarkable Achievements of Earthworms* by Amy Stewart
- *The Market Gardener* by Jean-Martin Fortier
- *Farmers of Forty Centuries: Organic Farming in China, Korea and Japan* by F. H. King
- *Resilient Agriculture: Cultivating Food Systems for a Changing Climate* by Laura Lengnick, PhD
- *Growing for Market* magazine growingformarket.com
- *Collapse* by Jared Diamond
- *The Urban Farmer* by Curtis Stone

Materials from Those Interviewed

- Seeds of Solidarity Farm in Orange, MA, holds workshops on their methods at seedsofsolidarity.org.

- Singing Frogs Farm in Sebastopol, CA, holds workshops and has some videos on Youtube. singingfrogsfarm.com
- Bare Mountain Farm has a lot of resources on their website baremtnfarm.com and Youtube.
- Shawn Jadrnicek of Wild Hope Farm wrote *The Bio-Integrated Farm*, which has info on his no-till and other innovative farming methods.
- Conor Crickmore of Neversink Farm mentors online courses, has a variety of resources, and also develops small farm tools at neversinktools.com.
- Jonathan and Megan Leiss of Spring Forth Farm have a page about no-till flower growing on their website springforthfarmnc.com.

Index

About the Author

Andrew Mefferd is editor of *Growing for Market* magazine. He has spent 15 years working on farms in six states, including a year working on a no-till research farm, and worked for seven years in the research department at Johnny's Selected Seeds. He travels around the world consulting with researchers and farmers on the best practices in greenhouse growing and sustainable agriculture. He is the author of *The Greenhouse and Hoophouse Grower's Handbook*, and has a passion for cooking and promoting local farming. He lives and farms in Cornville, Maine.

ABOUT NEW SOCIETY PUBLISHERS

New Society Publishers is an activist, solutions-oriented publisher focused on publishing books for a world of change. Our books offer tips, tools, and insights from leading experts in sustainable building, homesteading, climate change, environment, conscientious commerce, renewable energy, and more—positive solutions for troubled times.

We're proud to hold to the highest environmental and social standards of any publisher in North America. This is why some of our books might cost a little more. We think it's worth it!

- We print all our books in North America, never overseas

- All our books are printed on **100% post-consumer recycled paper**, processed chlorine-free, with low-VOC vegetable-based inks (since 2002)

- Our corporate structure is an innovative employee shareholder agreement, so we're one-third employee-owned (since 2015)

- We're carbon-neutral (since 2006)

- We're certified as a B Corporation (since 2016)

At New Society Publishers, we care deeply about *what* we publish—but also about *how* we do business.

Download our catalog at https://newsociety.com/Our-Catalog or for a printed copy please email info@newsocietypub.com or call 1-800-567-6772 ext 111.

New Society Publishers
ENVIRONMENTAL BENEFITS STATEMENT

For every 5,000 books printed, New Society saves the following resources:[1]

43	Trees
7,189	Pounds of Solid Waste
4,299	Gallons of Water
5,607	Kilowatt Hours of Electricity
7,103	Pounds of Greenhouse Gases
31	Pounds of HAPs, VOCs, and AOX Combined
11	Cubic Yards of Landfill Space

[1] Environmental benefits are calculated based on research done by the Environmental Defense Fund and other members of the Paper Task Force who study the environmental impacts of the paper industry.